D0053330

about the author

JOHN MEDINA IS a developmental molecular biologist and research consultant. He is an affiliate professor of bioengineering at the University of Washington School of Medicine. He was the founding director of two brain research institutes: the Brain Center for Applied Learning Research, at Seattle Pacific University, and the Talaris Research Institute, a nonprofit organization originally focused on how infants encode and process information. Medina lives in Seattle, Washington, with his wife and two boys.

VIDEO BONUSES

www.brainrules.net

Watch key principles on aging come to life

In his signature animated style, John Medina guides you through the effects of aging, plus what you can do for both your brain and your body.

Take a lively tour of the 12 original Brain Rules

This 45-minute film features John Medina on the 12 original Brain Rules for home, work, and school—from "Exercise boosts brain power" to "Sleep well, think well."

See Brain Rules parenting concepts in action

John Medina hosts fun videos on speaking in parentese, the cookie experiment, dealing with temper tantrums, and more. Plus, take our parenting quiz.

BRAIN RULES FOR AGING WELL

brain rules for aging well

10 Principles for Staying Vital, Happy, and Sharp

JOHN MEDINA

Pear Press

Requests for permission should be addressed to:

Pear Press
P.O. Box 70525
Seattle, WA 98127-0525
USA

This book may be purchased for educational, business,
or sales promotional use. For information, please visit
www.pearpress.com.

FIRST EDITION

Edited by Tracy Cutchlow
Cover designed by Nick Johnson

Library of Congress Cataloging-in-Publication Data
is available upon request.

ISBN-13: 978-0-9960326-7-4
10 9 8 7 6 5 4 3 2 1

Printed in the United States of America

To Sir David Attenborough,
role model and mentor-at-a-distance,
for the continual reminder that science doesn't make truces with truth.

contents

Body and Brain

Future Brain

10 brain rules for aging well

1.

Be a friend to others, and let others be a friend to you

2.

Cultivate an attitude of gratitude

3.

Mindfulness not only soothes but improves

4.

Remember, it's never too late to learn—or to teach

5.

Train your brain with video games

6.

Look for 10 signs before asking, "Do I have Alzheimer's?"

7.

MIND your meals and get moving

8.

For clear thinking, get enough (not too much) sleep

9.

You can't live forever, at least not yet

10.

Never retire, and be sure to reminisce

introduction

I PRESENT IN THE pages of this book everything you need to know about why you are aging. And I am going to use brain science to show how you can make life a surprisingly fulfilling experience—at least for your brain—in the years you have left. We begin with a group of seventy-year-old men in the capable hands of famed Harvard researcher Ellen Langer.

Lively—almost childlike—the seventy-year-old men skipped out of a monastery one fine morning. They'd just spent five days living in the old building, under observation by Langer. Now the men were leaving for home—smiling, happy, active, laughing. It was the fall of 1981, the first year of Ronald Reagan's administration, and the men had the same sunny abandonment associated with our fortieth president—who, coincidentally, was exactly their age. But these seniors, as part of Langer's research project, had just been through a time warp. Their brains had spent the past workweek not in 1981, but in 1959. The monastery was filled with songs like "Mack the Knife" and "The Battle of New Orleans." On the black-and-white TV, the Boston

Celtics beat the Minneapolis Lakers in the finals (yes, *Minneapolis* Lakers) and Johnny Unitas played for the *Baltimore* Colts. Issues of *Life* magazine and the *Saturday Evening Post* lay about. Ruth Handler had persuaded Mattel to create a thin, full-figured doll named after Ruth's daughter, Barbie, and then market it to little girls who had yet to undergo puberty. President Eisenhower had just signed into law the Hawaii Admission Act, creating the fiftieth state.

That walk down memory lane was the reason the men were so happy as they left the monastery. Waiting for the bus to take them home, a few entered into a spontaneous game of touch football—an activity most had not done for decades.

You might not have recognized these men 120 hours previously. They were shuffling, with poor vision, hearing, and memory; some of the men required canes to walk into the monastery. A few could not carry their suitcases up to their rooms. Langer and her team had poked and prodded the men's bodies and assessed their brains. These baseline tests proved one thing: before entering the monastery, the men were stereotypically old, as if ordered from Central Casting under the request "Eight infirm seniors, please."

But they didn't stay infirm. At the end of their stay, they underwent the same tests. Reading about the quantifiable change took my breath away. Even a casual visual inspection of these seniors revealed that something dramatic had happened, as the *New York Times* reported. Their posture was more robust. Their hands gripped more tightly. They handled objects with greater dexterity. They moved more easily (touch football, for heaven's sake!). Their hearing had sharpened. Same with their vision. Yes, *vision*. A sampling of their conversation would have told you something in their brains had dramatically improved too, and this impression would be proved by a second round of IQ and memory tests. In honor of its extraordinary finding, the experiment has been christened the "counterclockwise study."

The book you have in your hands is all about what happened to the men during those five days. And what will happen to you, statistically speaking, if you follow the advice in these pages. Such optimism is rare for me. I'm a grumpy neuroscientist. That means every scientific sentence in this book describes something published in the peer-reviewed literature, often replicated many times. (See www.brainrules.net/references.) I specialize in the genetics of psychiatric disorders. But if you think aging is all about debilitation, you may want to spend some quality time with another point of view, like Langer's. Or the one in this book.

Brain Rules for Aging Well describes not only how the brain ages but also how you can reduce the corrosive effects of aging. This field of inquiry is called geroscience.

As you peruse these pages, you'll discover what geroscientists already know. You'll learn how to improve your memory, why you should hang on to your friends for dear life—literally—and why you should go dancing with them as often as possible. You'll discover why reading a book several hours a day can actually add years to your life. You'll find that learning a new language may be the best thing for your mind, especially if you're worried about dementia. And that regularly engaging in friendly arguments with people who disagree with you is like taking a daily brain vitamin. You'll also learn why certain video games can actually improve your ability to solve problems.

Along the way, we'll dispel a few myths. Forget the double-your-order-if-you-call-now Elixir of the Fountain of Youth—there is no such thing. When it comes to causes of aging, wear and tear is less detrimental than a failure to repair. And it is *not* inevitable that your mind will power down as the years pass. If you follow the advice in this book, your brain can remain plastic, ready to study, ready to explore, and ready to learn at any age.

We'll also discover there are benefits to aging, with dividends paid not just to your head but to your heart. Your ability to notice the glass is half-full actually increases the older you get, and stress levels decline. That's why you should never listen to anyone who tells you old age is automatically filled with grumpy people. If you do it right, old age can be some of the happiest years of your life.

Four sections

Brain Rules for Aging Well is organized into four sections. First up, the social, or feeling, brain, exploring topics such as relationships, happiness, and gullibility to illustrate how our emotions change with age. Next, the thinking brain, explaining how various cognitive gadgets change with time. ("Gadgets" is my way of describing complex, interconnected brain regions with multiple functions.) Some actually improve, by the way. The third section is all about your body: how certain kinds of exercise, diets, and sleep can slow the decline of aging.

Each of these chapters is sprinkled with practical advice, explaining not only how certain interventions can improve performance but also what is known about the brain science behind each intervention.

The final section is about the future. Your future. It's filled with topics as joyful as retirement and as inevitable as death. I'll connect the previous chapters into a plan for maintaining your brain health. And you'll want to pay attention to all of them. The reason for this is nicely explained by the Amazon River. Or, rather, nicely explained by Sir David Attenborough's insights into the Amazon River.

A mighty river

As a youngster, I would watch the extraordinary TV documentaries narrated by this famed naturalist, and he disabused me of more errors about the natural world than I care to admit. One error had to do with the Amazon River.

I used to think the origin of the world's thickest river was a single burbling spring that somehow magically swelled in size as it flowed across the land. You know, like most rivers. I felt dismay when Attenborough pronounced that the Amazon had no such singularity. Like most rivers. Wading through a tiny stream in his *Living Planet* series, he intoned: "This is one of the many streams that can claim to be a source of the biggest river on earth—the Amazon!" And later: "The many sources of the Amazon began as numberless rivulets on the eastern flanks of the Andes." How disappointing! There was no single-origin story for 20 percent of the world's freshwater. There were many smaller sources, each making an *e pluribus unum* contribution to a final, massive outflow.

It's a pattern we'll encounter again and again. Take the memory chapter. Science shows that many factors contribute to keeping your massive memory streams flowing strong. Staying stress-free plays a role. So do regular aerobic exercise and how many books you read last week and how much pain you are currently experiencing and whether you get a good night's sleep. These factors serve as rivulets, each making a contribution to the larger Amazonian ability to recall things.

We now know that keeping the brain working well into old age involves creating lifestyles that act like streams high in the Andes. To best understand how we can retain our own intellectual effervescence, this book will wade into the contributions of each stream.

Toward the end of our discussion, I'll describe how scientists are trying to hack into the molecular machinery of the aging process itself, tinkering with its "inevitability code" in an attempt to reverse the irreversible. As an AARP-eligible father, I embrace this effort wholeheartedly, though as an AARP-eligible scientist, I temper my enthusiasm with a healthy dose of scientific grumpiness.

It will then be time to revisit Langer's lively septuagenarians, for the results of her time-warp studies will now make more sense. I won't be sugarcoating the harsh ways in which time can run roughshod over the human experience. But you will come away understanding that there is a lot more to aging than aches and pains and longing to return to the days of the Eisenhower administration.

It's a good time to grow old

We've got it relatively good. For virtually our species' entire history, human life expectancy was about thirty years. Life expectancy is the benchmark for what's typical. And it has been steadily rising. Were you living in England in 1850, you generally died in your mid-forties. That figure is four decades longer now. If you were an American in 1900, you died around age forty-nine. It was seventy-six by 1997.

Not true anymore. Americans born in 2015 can expect to live to seventy-eight (it's a little more for women, a little less for men). If you've already made it to your sixty-fifth birthday, you can expect to live nearly twenty-four more years if female and nearly twenty-two more years if male. That's an astonishing 10 percent jump since the year 2000, and the numbers are expected to go even higher.

If life expectancy gives us a benchmark for what's typical, what's possible?

When we look at the years a creature is capable of living, we're talking about *longevity* (more properly, longevity determination). This number is regulated, somewhat indirectly, by genes. If you used the term "genetic longevity determination," researchers in the room would nod their heads in approval.

This notion is different from *maximum life span*, and both are different from *life expectancy*. It's easy to conflate them, which would earn you a frown from those researchers. The scientific journal *Nature* published succinct definitions a few years back: "Maximum life span is a bald measure of years accumulated. It is not the same as life expectancy, which is an actuarial measure of how long one is expected to live from birth, or indeed from any given age."

In this view, longevity is the amount of time you could spend on the planet were conditions ideal. Life expectancy is the amount of time you likely will spend on the planet, given that conditions are almost never ideal. It's the difference between how long you *can* live versus how long you *will* live.

So how long can humans live? The oldest person with an independently verifiable birth date celebrated her 122nd party before passing. But most of the oldest people clock in between 115 and 120 years old. You'd have to weather a lot of biological perfect storms to get to your 120th birthday party, of course, and almost none of us will. The probability isn't zero, though.

We really are learning how to soldier on right to the edge of our expiration dates. And, as the stories throughout this book illustrate, we're doing it in greater physical and mental health than at any other time in our history.

But these stories can't tell you how *you* will age. That's because aging is quite variable—even individually expressed. There's an intricate fox-trot between nature and nurture. And the fact that the brain is so flexible, so damnably reactive to its environment, is actually a powerful confounder for many types of brain research. The brain appears hardwired not to be hardwired. Consider the simple act of reading this sentence and discovering I've left the period off the end of it The very fact that I did, and that I told you, and that you probably looked to see if I was telling the truth, *physically* rewired your brain.

How the brain is wired

Whenever the brain learns something, connections between neurons change. What does that look like? Neural circuitry has many options. Sometimes the changes involve neurons growing new connections to the locals. Sometimes the changes involve abandoning certain connections and re-forming new ones somewhere else. Sometimes the alterations only involve electrical relationships between two neurons, something called synaptic strength.

You probably learned in high school that brains are strung together with electrically active nerve cells—neurons—but you may have forgotten what they looked like. To illustrate, I'd like to introduce you to what are easily the First Ladies of my wife's garden, our two graceful Japanese maples. They're beautiful creatures, more bush than tree, with elegant, tapered leaves, deeply red in the autumn. These leaves are fastened to complex branches, which gather at a stubby trunk. The trunk is nearly hidden from view, given the exuberance of the branching, and the little you can see quickly dives under the soil.

The underground part of the maple splits into a slightly less complex root system, like most plants.

Though neurons come in many shapes and sizes, all follow a basic structure, looking something like our garden's Grand Dames. Impossibly complex branching structures, called dendrites, exist at one end of a typical cell. Those dendrites gather together into a trunk-like structure termed an axon. Unlike our maple's trunk, however, there is a bulge at this point of gathering. It's an important swelling—called the cell body—and its reputation derives from a small spherical shape inside it. This is the nucleus of the neuron. It houses the cell's command and control structures, the double-ladder-shaped molecule DNA.

Axons can be short and squatty, like our maple's trunk, or long and slender like a pine tree's trunk. Many are wrapped in a type of "bark" that's called white matter. At the other end of the axon lies a root system, just like a plant's, consisting of branching structures termed telodendria. These usually aren't as complex as the dendrites, but they serve an important information-transfer function, as we're about to see.

The brain's information system runs on electricity, like most light bulbs, and their shape helps them do it. To understand how, imagine pulling one of our Japanese maples out by its roots, and then, while my wife has a heart attack, holding it over the top of our other maple. Don't let them touch. The root system of the top tree is now hovering over the branches of the bottom.

Now imagine these two trees are neurons. The telodendria (roots) of the upper neuron lie close to the dendrites (branches) of the lower cell. In the real world of the brain, electricity flows from the dendrites of the top neuron down its axon and arriving at the telodendria, where it immediately encounters the space between the two. The gap must be jumped if information is to be transferred. This junction is called a synapse, and the space it creates, the synaptic cleft. How to pole-vault the space?

The solution lies at the tips of those root-like telodendria. There are small bead-like packets at those tips containing some of the most famous molecules in all of neuroscience. They're called neurotransmitters. I'll bet you've heard of some of them: dopamine, glutamate, serotonin.

When an electrical signal reaches the telodendria of one neuron, some of these biochemical celebrities are released into the synaptic cleft. It's the equivalent of saying, "I need to send a message to the other side." The neurotransmitters dutifully sail across the gulf. It's not a long journey; most of these spaces are only about 20 nanometers in length. Once the neurotransmitters have crossed, they bind to receptors on the dendrites of the other neuron, like a boat tying up to a dock. This binding is sensed by the cell, alerting it with signal that says: "Oh, I better do something." In many cases, that "do something" means becoming electrically excited too. It then passes along this excitement down the chain from dendrites to axons to its telodendria.

While jumping the space between two neurons using biochemicals is a neat trick, the electrical circuits aren't usually this simple. If you can imagine lining up thousands of cellular Japanese maples root-to-branch, you'd have something approximating an elementary neural circuit in the brain. And even that's too simple. The typical number of connections a single neuron makes with other neurons is around seven thousand. (That's only an average: some have more than a hundred thousand!) Under the microscope, neural tissue looks like thousands of maple trees have crashed together in one space, whipped by an F5 tornado.

These are the structures that change so flexibly when the brain learns something new. These are the structures that become damaged as we age. However, there's another fascinating reason that the damage of aging is incredibly individual.

The brain doesn't just react to changes in the outside environment. Remarkably, the brain can respond to changes it observes happening to *itself*. How does it do that? We've no idea. We do know that if it senses the changes are likely to be negative, it can create work-arounds to fix the problem.

Cells erode, lose connections, or simply stop functioning. These alterations could easily lead to behavioral changes, but they don't always. The reason is that the brain kicks into compensatory overdrive and reroutes itself according to a new plan.

The major culprit in aging is a hot topic. Some scientists speculate about immune system deficiency (the immunologic theory). Others blame dysfunctional energy systems (the free radical hypothesis; mitochondrial theory). Others point to systemic inflammation. Who is correct? The answer is all of them. Or none of them. Each hypothesis has been found to explain only certain aspects of aging. The sum total is that many systems get hit as we grow old, but which ones sign off first is individually experienced.

There are nearly as many ways to transit through the aging process as there are people on the planet. It's a theme as familiar as shopping for jeans: one size does not fit all. Discernible generalizable patterns do exist, and studying the brain is a great way to see some of them. But to get an accurate view, we're going to have to gaze at an occasionally cloudy statistical mirror. It's okay. We'll still look fabulous. We'll just be a little older.

Our goal is to learn how to create lifestyles that will continually grease the biological gears controlling how long we live. And how *well* we live. Fortunately for us, geroscience is well funded. Scientists have discovered many cool things we can do as our brains age. All of these discoveries over the years add up to one thing: science is literally changing our minds about the optimal care and feeding of the brain. All of it is captivating. A great deal of it is unexpected. One of the most delightful is the subject of our first chapter. It's the jovial power of having lots of friends.

SUMMARY

- Geroscience is the field of inquiry dedicated to studying how we age, what causes us to age, and how we can reduce the corrosive effects of aging.
- Aging is mostly due to the breakdown of our biological maintenance departments, our body's increasing inability to repair the day-to-day wear and tear adequately.
- Today, we humans are living much longer than we have for the majority of our existence. We are the only species capable of living past our prime.
- The human brain is so adaptable that it reacts to changes not only in its environment but also within itself. Your aging brain is capable of compensating for breakdowns in its own systems as you get older.

SOCIAL BRAIN

your friendships

brain rule

Be a friend to others,
and let others be a friend to you

My favorite kind of pain is in my stomach
when my friends make me laugh too hard.
—Anonymous

At some point, you have to realize that some people
can stay in your heart but not in your life.
—Sandi Lynn, author of Forever Black

HERE'S A SENTENCE YOU probably don't want to hear from Dad an hour after your wedding: "I'll tell you what. If it lasts more than a year, I'll give you a hundred bucks."

Unfortunately, that's exactly what happened to Karl Gfatter, a story he enthusiastically relates in a nursing home, wheelchair bound now, his loving bride at his side. And Dad had to pay up, probably many times over, for Karl and Elizabeth have stayed together for more than seven decades. Karl related this comment to the local media, who dropped by as he and Elizabeth were celebrating a recommitment ceremony in honor of their seventy-fifth wedding anniversary. They were surrounded by residents, staff, clergy. And rice. Plus lots of joy, smiles, and even some tears, creating the feeling you'd just walked onto the set of *It's a Wonderful Life*. Both were radiant, bright as buttons. "We eloped because they didn't want us to get married yet. They said we were too young!" Elizabeth laughed.

What Karl and Elizabeth may not know is that having a long marriage—and a room full of friends—is helping to keep their brains young. Friendships, and the social activities that surround them, are the major focus of this chapter. We'll discuss the cognitive power of maintaining friendships over many years, along with the opposite: loneliness. Then we'll dance our way toward a surprisingly beneficial brain booster.

Socializing: vitamins for the brain

You'd have a hard time finding someone more socially active—and intellectually lively—than wealthy heiress and arts patron Brooke Astor. By the year 2000, she was New York royalty, married to a man whose father actually died on the *Titanic*. Along with three of her closest friends—fashion publicist Eleanor Lambert, former opera singer Kitty Carlisle, and fashion designer Pauline Trigère—Brooke tore through a social schedule that required four changes of clothing a day. Lunch at a downtown café, then a board meeting at the Museum of Modern Art (she was a trustee), an evening concert at Carnegie, followed by a benefit dinner, ending with late drinks, returning home in a comet tail of paparazzi flashbulbs.

Brooke kept a social schedule that could leave a twentysomething personal secretary exhausted. And did—which is in great contrast to the physical ages of the women in this smart, lively quartet. Kitty, the youngest of the bunch, turned ninety that year. Pauline was ninety-one; Eleanor, ninety-six. Brooke was ninety-eight years old.

Did their age, social activity, and intellectual vigor have anything to do with one another? The answer, to the acclaim of elderly partygoers everywhere, is yes. Social interactions are like vitamins and minerals for aging brains, with ridiculously powerful implications. Even socializing over the Internet provides benefits.

The studies are anchored in the safe harbor of peer-reviewed research. The first set of studies established a solid correlation between social interactions and cognition. Researcher Bryan James, an epidemiologist with the Rush Alzheimer's Disease Center, assessed the typical cognitive function and social interactivity of 1,140 seniors without dementia. He scored their social interactivity, then measured their rate of global cognitive decline over a twelve-year period. For the group that socialized the most, the rate of cognitive decline was 70 percent less than for those who socialized the least.

Other researchers focused on specific types of cognition and found virtually the same thing. One famous study looked at rates of memory decline in social isolates versus social butterflies, examining a staggering 16,600 people over six years. Memory decline of the Brooke Astors was half that of the shut-ins. A flurry of other findings confirmed a robust correlation between social interactions and cognitive health.

Even better, the next set of studies looked at causation, not just correlation. They measured people's baseline cognition, introduced some form of socialization, then remeasured cognition. One intervention showed a cognitive boost in processing speed and working memory with as little as ten minutes of social interaction. Like a public television fund-raiser, data linking socialization with brain power turns out to be remarkably persistent.

The interactions don't have to be within a long-term relationship, and they don't necessarily refer to the number of friends one has. Researchers who study this stuff use words like "positive social interactions" (generally associated with the release of dopamine in the brain), "negative social interactions" (generally associated with hormones such as catecholamines and glucocorticoids, released in response to stress), and "social exchanges" (to describe interactivity). I'm going to use the word "relationships" more often to keep things friendly. But if you have social interactions that are positive—whether deep or momentary, with one person or dozens—benefits accrue.

What about the digital world? Does the social interaction have to be in person? Researchers realized long ago that the Internet might provide a perfect way for socially isolated, mobility-challenged seniors to interact with others. The rise of video chats created a terrific experimental test bed. Could people increasingly tethered to home still get a brain lift?

The answer, welcome as a Rothko retrospective, was again yes. One experiment involved people eighty years and older, measuring a baseline for executive function skills and an aspect of language ability that's related to executive function. Executive function (EF) is a behavioral gearbox mostly housed in the prefrontal cortex, an important region located right behind your forehead. EF includes cognitive control (such as the ability to shift attentional states), emotional regulation (such as the ability to manage your anger), and short-term memory. The researchers got baseline EF scores, then installed a video-chat program for each person and proceeded to hold conversations with the octogenarians, averaging thirty minutes per day for six weeks. Four and a half months later, their brains were retested.

Researchers observed large improvements in both executive function and language skills. The scores leapfrogged over controls who spoke for thirty minutes by phone only. This is consistent with other data suggesting that the better you simulate actual human contact, the richer the social experience becomes. Video chat is not perfect, but for those without the option of regular human contact, it's a godsend.

These findings are worthy of a J. D. Power award for senior citizen customer satisfaction. Which means you should get out your social calendar, iron your best clothes, and go run a board meeting. Or visit a museum. The answer to the question "Does socialization really decrease the rate of cognitive decline?" is a robust and hearty "Yes."

How exactly does the buoyant power of socialization work? Two main ways: it reduces stress, which helps maintain not only the body's general health but specific aspects of the immune system, and it's a workout for the brain.

More parties, less flu

The more positive social interactions you have, the lighter your allostatic load becomes, as neuroendocrinologist Bruce McEwen would put it. He's the researcher who came up with the concept of "allostatic load." Allostatic load is the aggregate effect of stress on your body's capability, including brain capability, through time. The more stress you encounter, the bigger the load (and the greater the damage). Consider stress metaphorically: the stresses in life are oceanic waves, and your body is a cliff. The more waves that crash onto the cliff, the greater the erosion, and the more severe the total effect. Allostatic load is the measure of your body's deterioration, in response to the lifelong waves of stress you experience.

Less stress is important particularly for the immune system. The immune system naturally becomes compromised as you age, but the more stressed you are, the greater risk you run of weakening parts of the immune system. We even know why. One critical arm of the immune system involves a group of cellular warriors known as T-cells. These cells play critical roles in wound healing (like when you get a cut) and recovering from infectious diseases (like when you get colds and flus). Stress hormones like cortisol—at the high levels you experience when you're in a bad marriage or otherwise chronically stressed—actually kill T-cells. Your wounds heal at a rate 40 percent slower if you're in a high-hostility marriage than in a low-hostility one. And you get more colds. Says elderly-care expert Gary Skole: "Those elderly folks who get out and interact and spend more time with people during cold/flu season actually get fewer colds and illnesses than those who spend most of their time alone."

These data serve to underscore the growing link in the scientific literature between positive interactions, stress reduction, and longer life. No doubt Karl and Elizabeth are right now busy nodding their heads. And Karl's dad is probably rolling around in his grave.

A workout for your brain

One of the reasons why social interactions are so good for you is that they take so much energy to maintain, consistently giving your brain a bona fide workout. Case in point is a clip from the movie *When*

Harry Met Sally. The scene is where Sally (Meg Ryan) asks Harry (Billy Crystal) to come over for some major-league consolation: Sally's ex has decided to marry someone else. Through tears and sobs and gobs of tissues, Sally tells Harry, "All this time, I've been saying that he didn't want to get married. But the truth is, he didn't want to marry *me*." Harry, bless him, attempts his best lifeboat impression, although by now Sally is nearly drowning in a cocktail of saltwater and snot. "I'm difficult!" she blubbers. Harry counters thoughtfully: "Challenging." Sally sobs, "I'm too structured, I'm completely closed off!" Harry shrugs: "But in a good way."

With unexpurgated grief in Sally's case and measured restraint for Harry, the amount of energy the two exude in this delightful scene is extraordinary. It illustrates something scientists have known for years: flesh-and-blood friendships take *work*. And that's because social interactions take *work*. And by *work,* I mean in a biochemical, energy-expending kind of way. Some researchers believe social interactions are the most complex, energy-intensive jobs your brain can consciously perform. Every time it intermingles at a cocktail party or consoles a friend, the organ experiences the cognitive equivalent of an aerobic workout.

Says Chelsea Wald, writing in *Nature* magazine: "[Researchers] suspect that the cognitively demanding act of socializing can actually build up the brain—like exercising builds up muscles. This 'brain reserve' may then act as a buffer against functional loss, even in the face of conditions such as Alzheimer's disease."

Suppose you were the scientist hypothesizing that social interactions are cognitive calisthenics. You might predict that the more social interactions you have, the more you exercise the brain regions responsible for those interactions. You might further hypothesize that the neural tissue will become bigger and stronger or more active as a result. You might guess there would even be bleed-through effects, given that the job descriptions of most brain regions are hopelessly intertwined with those of other regions, all moonlighting to produce a broad array of functions. From cell to behavior, you can measure whether growth is occurring.

And scientists have. Though the data are largely correlative, growth is exactly what they find.

Let me pause for a moment to define a few terms: social activities, social networks, and social cognitions. Researchers define these terms much as the public does, especially if that public uses words like "neurological substrates." Social activities are the actual experiences you have with others, whether going out on a boat or going out on a date. Social networks are the number of people with whom you willingly have those experiences. Close friends and family generally populate these activities. Social cognitions are the psychological (and by implication, neurological) substrates you use to interact with others when socializing.

On to the studies showing that the brain is being exercised.

The more social relationships you maintain, the bigger the gray matter volume in specific regions of your frontal lobe. Which means that relationships are to the frontal lobe what milk shakes are to your waistline. The frontal lobe is the large region right behind your eyes, running to the middle of your head (where a headband would sit). This region is associated with a cognitive gadget called mentalizing, or Theory of Mind. Mentalizing is the ability to discern the mental states of others, particularly their motivations and intentions. It's as close to mind reading as your brain will ever get. Mentalizing abilities play a powerful role in establishing and maintaining social relationships, as you can imagine.

The frontal lobe is also responsible for helping you predict the consequences of your own actions. It helps you suppress socially inappropriate behaviors and even make comparative decisions. For many reasons, these are important regions to keep fat and happy.

The amygdala, a little almond-shaped nodule dangling just behind each ear, is involved in processing your emotions. It too is affected by levels of social activity. The higher the overall number of (and the greater the variability in) the types of relationships you maintain, the bigger your amygdala becomes. These aren't small changes. If you triple the number of people in your social network, you double the volume of your amygdala. Wondering how you'd keep up with all those people? While you maintain your closest relationships with five people at a time, researchers find, you can have meaningful relationships of varying quality with an additional 150 people. Think of it as rings of relationships.

Social activity also affects a region called the entorhinal cortex, which helps you recall important things like your first kiss. This romantic bundle of nerves, which also helps process other types of memories (and many types of social perceptions), is located in the temporal lobe, the brain regions closest to your eardrums.

Given the rise of the Internet, does it matter which kind of social network is being measured, silicon- or carbon-based? It does. For example, gray matter changes in non-amygdalar regions (like the frontal lobe and entorhinal cortex) occur only with flesh-and-blood interactions. In contrast, density changes in the amygdala are specifically associated with the size of both Web-based social networks *and* the number of face-to-face social interactions. The reasons for these differences, extraordinary as they may sound, are not known.

Not all social interactions are created equal, however. You don't have to look any further than a typical day in an American office, populated by dysfunctional management, for an example.

The boss from hell

The boss wore his unpleasantness like a purity ring on his middle finger. He publicly announced the contents of private meetings to his entire forty-person staff. He slapped the hand of a loyal employee who had worked for the company for forty-four years. When that employee asked for time off to go to the hospital where her daughter had suddenly been admitted, the boss replied, "What are you going to do, hold her hand?"

I describe this narrative, one of many stories online chronicling chronically bad working relationships, to counter an impression you might be getting from this chapter: that every relationship provides neurological benefit. The truth is just the opposite. You can have many relationships with people, but if they're negative, they're unhealthy. Studies show that it's not the overall number of interactions that benefit health, but the net quality of the individual interactions. According to researchers from the University of North Carolina at Chapel Hill: "Social support and strain, which measured qualitative characteristics of social connections that are distinct from relationship quantity, mattered more for physical health in mid-adulthood, and continued to have impacts in late adulthood."

Behavioral labs are coming up with all kinds of dos and don'ts for relationships. Interactions burdened with competitive one-upmanship provide no cognitive benefit at all. Relationships with people who are emotionally controlling, meddlesome, or consistently verbally aggressive (like that aforementioned boss) are worth limiting, if not ending altogether.

Drop the ego

What's the secret to a good interaction for your brain? It's a willingness to consistently take the other person's point of view, actively seeking to understand a different perspective. You may agree with the other person or you may not, but the effort transforms casual conversation into meaningful brain food. If that sounds like Theory of Mind stuff we've been talking about, you are right on the research money. It's also a scientifically nice way of saying: stop being so self-centered. This advice, by the way, is just as healthy for people much younger than your average Social Security recipient. Regularly engage people, and your brain will thank you at any age.

You can create an environment conducive to quality relationships. Social psychologist Rebecca Adams summarized how in a *New York Times* interview a few years back, if you cultivate the following:

- "repeated, unplanned interactions," spontaneously rubbing shoulders with good friends
- "proximity," living close by to friends and family members so those shoulders are available for rubbing
- "a setting that encourages people to let their guard down"

Not surprisingly, Adams relates, most of our tightest friendships initially form in college, where these conditions are met by design.

It's best to have friends of all ages—including kids. That notion may transcend our culture's perspective, but not our culture's data. The more intergenerational relationships older people form, the higher the brain benefit turns out to be, especially when seniors interact with elementary-age children. It reduces stress, decreases rates of affective disorders such as anxiety and depression, and even lowers mortality rates.

There are probably many reasons for these findings. Young people always have different perspectives from their elders. That means regular exposure to virtually anyone of a different generation increases the diversity of opinions you're likely to experience. The music to which you listen may change. You may read different kinds of books, learn to laugh at different things. If you regularly inhabit another's point of view, you are exercising very important regions of the brain. The quote "Sometimes you need to talk to a three-year-old so you can understand life again" is quite literally true. Plus, if the only friends you have are old, you will be attending many more funerals than weddings. And there's nothing like watching the death of people around you to increase your sense of isolation. Having younger friends opens up a healthy can of life-goes-on, with a sparkling supply of weddings and baby showers in case you forget. Statistically, you've got a guarantee your young friends will outlive you.

Happily, the benefits of intergenerational friendship flow back into the life of the child. Regular interactions with older people increase a child's problem-solving skills, positively influence emotional development, and improve language acquisition. Older people tend to be more patient, tend to look on the sunny side of life, and are more experienced with kids, often having raised children of their own. This ability to be kind, to listen, to empathize, is especially valuable for kids being raised in the chaos of a two-career family. Kids may always be the demander-in-chief, yet seniors who can make time for them and all their youthful foibles will discover the joys of being a wiser parent this time around.

So become someone's favorite grandparent, as well as a mentor, friend, and confidant. Create peace in your marriage. Make friends with your neighbors. See your friends often.

And if you don't?

All the lonely people

Researchers have uncovered three important facts about old age and loneliness. The first is as welcome as wrinkles: loneliness really increases with age. Depending on the study, the proportion of older adults experiencing at least moderate amounts of loneliness is anywhere between 20 percent and 40 percent. Second, loneliness throughout a person's lifetime is uneven, following a U-shaped curve. Third, loneliness is the single greatest risk factor for clinical depression.

The definition of loneliness seems as obvious as drywall. You want to be around people and you can't, so you feel bad. Defining loneliness in a scientifically specific way, though, is a bit tricky. Some people are "loners" and prefer life that way. Some folks favor pets over people. Others need humans around all the time. Researchers use the term "objective social isolation" for those who are isolated (and may even prefer it) and "perceived social isolation" for those who feel alone (and definitely do not prefer it). Here's a laboratory definition for you: "A perceived lack of control over the quantity and especially the quality of one's social activity."

Scientists also have a psychometric test to measure what that quote means. Developed in one of the least lonely places on earth, Southern

California, the test is appropriately called the UCLA Loneliness Scale. Here's what researchers have found.

We start feeling lonely in late adolescence, and the feeling decreases as we move through early-to-middle adulthood. That's natural: we go through school, jobs, kids—experiences chock-full of other people. Our number of friends rises sharply to peak at age twenty-five, then slowly drifts down to age forty-five, levels a bit, and continues its decline after fifty-five, completing the U shape of loneliness.

There are many caveats and nuances to these data, so the U curve's a bit wobbly. Seventy-five-year-olds experience some of the *least* feelings of loneliness in life, followed by the *most* a month or two after their eightieth birthday. Seniors who don't make much money experience severe loneliness more sharply than seniors who do: a monstrous threefold increase. Married people experience less loneliness than those living alone. This is true for all age groups, but the quality of intimacy plays a larger role for the marital well-being of seniors than of younger people. Physical health plays a powerful role in how much isolation the elderly suffer, too.

Where social isolation leads

The more socially isolated you become, the less happy you are. Researchers believe the reasons for this are deeply rooted in evolution: humans were too weak, biologically speaking, to survive without each other for long. Our brains created a system of negative responses to social isolation, compelling us to seek each other out. Cooperation and the mentalizing tools we developed for it put us squarely into the Darwinian carpool lane. We then survived long enough to pass along our genes.

We don't do very well when we get lonely. For one, our social behaviors begin eroding. Loneliness is associated with poorer grooming habits, for example, and an increasing inability to navigate intimate life functions such as bathing, using the toilet, eating, dressing independently, and getting out of bed. Some of this may be related to the oncoming squalls of depression, gusts to which lonely seniors are particularly vulnerable.

Lonely seniors have poorer immune function. They can't fight off viral infections or cancers as easily. They have higher levels of stress hormones, which bring on all kinds of negative effects. Chief among these are higher blood pressure, which increases the risk for heart disease and stroke. Loneliness hurts overall cognition too, from memory to perceptual speed. It's even a risk factor for dementia.

Chronic loneliness can throw you into a nasty loop. As you probably know, the process of aging involves physical pain: certain tissues begin to break down for which there will be no cure; aches intensify in specific body parts naturally vulnerable to aging (arthritis is but one example). Such discomfort can affect your topics of conversation, your mobility, and your sleep. All combine to make you increasingly unpleasant to be around. The more unpleasant you are, the less people want to hang with you. Fewer social interactions make you more susceptible to the problems we've been discussing. You become even more unable to interact socially, and people quit visiting. This cycle repeats itself over and over again: the lonelier you are, the lonelier you become. And that's when the attack dog of depression strikes. By the time people are in their eighties, loneliness is the single greatest risk factor for clinical depression. That's a steaming bag of bad neural news, as we'll discuss in a later chapter.

The most dramatic effect of social isolation on the elderly is death. The probability of death is 45 percent greater for lonely seniors than it is for socially active ones. That number holds steady even when you control for things like debilitating physical ailments and depression. If you don't have a lot of friends, you die sooner than you have to.

Inflammation of the brain

"Tell us, Mrs. Holderness, what do you think is the best thing about being 103?" a journalist asked. Molly's response was quick and good-humored: "No peer pressure."

She is fortunate to have a sharp mind. Many elderly people don't—and most of those are women. Neuroscientist Laura Fratiglioni wondered if there could be a connection between the fact that men die before women, leaving widows alone in life, and the fact that women suffer more dementia than men, especially after the age of eighty. Could isolation be the culprit? Fratiglioni determined there was indeed a correlation. Women who live alone, as well as those without strong social interactivity, are at much greater risk for dementia than those who live with someone or have sustained, close social interactions.

The brain mechanisms behind this disturbing finding were soon under active investigation. A clear, more causal picture has emerged: excessive loneliness causes brain damage.

This deserves a fuller explanation because it's a really big thing to say. The biological machinery involves, of all things, the same mechanisms stimulated when you stub your toe.

You undoubtedly know about inflammation. You stub your toe and local infectious agents—like bacteria—sweep in to take advantage, launching their Lilliputian attacks. Your body responds with swelling, redness, profanity. The classic inflammatory response is supervised by many molecules, including ones called cytokines. The response usually doesn't last very long; the cytokines do their job and, in a few days, destroy the unwanted bad actors. This is a case of acute inflammation.

There is another type of inflammation, however, related to stubbed toes and also involving cytokines, but more relevant to our story. It is called systemic or persistent inflammation, the key difference tucked into its name: it lasts a long time. This type of inflammation occurs all over the body. It's akin to getting tiny toe-stubs throughout the major organ systems, then having your whole body react with systemic, low-intensity inflammation as a result.

Don't let the phrase "low intensity" fool you. Systemic inflammation damages many types of tissue over a long period of time, the way acid rain eats into a forest. It can even damage the brain, particularly white matter. White matter is composed of myelin sheaths that wrap around neurons, providing insulation to improve electrical performance. Without it, the brain doesn't function very well.

How do you get systemic inflammation? The paths are many, including environmental factors such as smoking, exposure to pollution, or being overweight. Stress, ever the acid reflux of behavior, can incite it. And so can loneliness, according to Timothy Verstynen, director of the Cognitive Axon Lab at Carnegie Mellon University. He found in 2015 that chronic social isolation increases the level of systemic inflammation. Just how much damage loneliness causes in humans turns out to be astonishing. It's at the same level as smoking. Or being too fat. The proposed molecular mechanism for this extraordinary observation is like a three-step feedback loop from geriatric hell: (1) loneliness causes systemic inflammation, (2) the inflammation damages white matter in the brain, and (3) the damage leads to the changes in behavior we mentioned, the ones resulting in fewer social interactions. Repeat.

If there is that thin a membrane between loneliness and brain damage, we have some serious thinking to do about how society treats its seniors. And how seniors treat themselves. We need to spend some quality time being grateful for the friends we have, and if the friendship tank is low, we need to seriously strategize about how to refill it.

A cultural shift

Refilling your friendship tank can be tough to do as you age. Researchers know you increase the quantity of friends you have in life until about age twenty-five. Then the number begins a long, slow decline, a deterioration that won't stop until late into middle age. Baby boomers are notorious for losing friends in later life. As seniors, they have fewer social interactions with people of nearly every stripe—family members, friends, next-door neighbors—than seniors did in the previous generation.

Sociologists concur there are multiple reasons for this decrease, though not every researcher agrees on exactly what they are. Some point to the fact that people of child-bearing age move around a lot. This means communities are constantly being formed, uprooted, and re-formed—not a condition conducive to creating rich, long-lasting adult friendships. As a result, the guarantee of relational stability that comes from staying in one place gets torn up. My grandparents celebrated the multi-decade wedding anniversaries of friends with whom they had also shared a first-grade classroom. Such a thing is almost beyond imagining today.

It doesn't help that people in developed countries are having fewer children than a generation ago. Over time, this means fewer uncles, aunts, and cousins. Even though that also means fewer annoying family reunions to attend, it shrinks the probability of sustaining long-term relationships with relatives (even if you did stay in one place). So you don't have close friends. You don't have much family. You barely even have a home. In terms of breeding toxic isolation, that's like stagnant water to a mosquito.

On top of that, the nature of friendship is changing. The digital world provides enticing electronic substitutes for flesh-and-blood interactions. An intense research effort is under way to see if this matters, and I'll have more to say about it in a later chapter.

The bottom line: environmental forces put seniors at greater risk for being alone than ever before. That's noxious, for at a time when your brain is already under corrosive assault from Darwinian-approved natural causes, social isolation is the last thing it needs.

And that's not even the full story. Nature plays just as strong a role as nurture. It is to these ideas that we next turn.

Face time

Prosopagnosia. It's tough to pronounce, tougher to experience. People who suffer from the P-word aren't able to do something even infants can do: recognize faces. They may have known you for years, but they won't recognize you if you walk into the room five minutes from now. Nor will they recognize anybody else, even though they usually can recognize every *thing* else. No problem with hats, for example, or with eyebrows, or even with the concept of "face."

Sufferers of prosopagnosia (logically called face blindness) usually resort to extraordinary measures to navigate their social world. A person might have to memorize the clothes their family members regularly wear in order to tell them apart. Others might have to pay close attention to the way people move or to specific postures, in order to recognize people at work. The late neurologist Oliver Sacks, a famous sufferer of face blindness, would have his guests wear name tags at parties so he could recognize them.

Not surprisingly, many people with the disorder withdraw socially, often suffering from social anxiety. This makes a certain amount of sense, for a great deal of social information is carried by the face. Clues to whether someone is happy or sad, contented or disgusted, potential mate or potential threat, show up in the eyes, cheeks, and jowls. Without the knowledge of what someone is feeling, sufferers withdraw into a *Twilight Zone* world where people can recognize you but you can't return the favor. Sacks himself quit attending conferences and large parties.

Prosopagnosia is associated with lesions in a brain region called the fusiform gyrus, an area in the lower part of your brain not far from where your spinal column enters your skull. Strokes and various head traumas can damage the fusiform gyrus. Face blindness also is as heritable as eye color, which means you can get it from your parents. It is thought to affect about 2 percent of the population. But a less severe form of it seems to be related to normal aging as well.

As people get older, they suffer an increasing inability to recognize familiar faces, and they lose their perception of some of the emotional information those faces carry. We even know the reason. The neural tracts—the white-matter cabling—connecting the fusiform gyrus to other regions of the brain begin to lose structural integrity. Prosopagnosia illustrates an important principle in the brain sciences: specific regions of the brain exert a dictatorship over specific functions. When those regions become injured, those functions can be altered—or disappear.

The behavioral deficits are not global. Seniors can recognize emotions like surprise, happiness, and even disgust just fine (in fact, they score better on tests measuring disgust than younger adults do). Not so with sadness, fear, and anger. It's an unfortunate twofer: seniors have a harder time recognizing people they know, sort of like a mini-prosopagnosia, and they have a harder time recognizing certain feelings those people are experiencing.

Do seniors withdraw socially as a result of these deficits, similarly to people with face blindness? Though there is (always) the need for further research, the answer may be yes. As we discussed, people begin to withdraw from social interactions as they age (remember the peak at twenty-five and downward slope at fifty-five?). Seniors show an especially severe reduction. Interestingly, the same shrinkage in social activity occurs in lab-raised monkeys when they become elderly.

We've talked about mentalizing, or Theory of Mind. As you get older, the ability to mentalize begins to decline. In a lab assay called the "false belief task," people try to guess the intention of someone else. Younger adults routinely get the correct answer about 95 percent of the time, elderly adults about 85 percent of the time. The senior scores worsen with age, such that after age eighty, the scores shrink to less than 70 percent. The reason appears to be an age-related change in the functional activity of a single region in the prefrontal cortex. The prefrontal cortex (often abbreviated as PFC) is evolution's newest add-on to your brain's fundamental architecture. It's a most talented structure, with functions ranging from decision making to personality formation. As we'll discover later, most of the talents we identify as uniquely human arise in the PFC.

Is it possible that changes in facial recognition and changes in mentalizing ability are related? And if so, might they be part of nature's contribution to the social isolation experienced by many of our elderly? The real answer is we don't know. But the fact that

I can write about this stuff in a scientifically meaningful fashion at all represents a tremendous leap in our understanding from even a few years ago. Such progress has even bled into the practical realm of intervention. Solid research shows steps we can take to ameliorate the negative effects of loneliness. It is to these steps that we turn next.

Dance the night away

The years of age separating dancers Mikhail Baryshnikov and Fred Astaire span about a half century. No matter: the Latvian's admiration for his American colleague is evident. "No dancer can watch Fred Astaire and not know that we all should have been in another business," said the legendary Soviet and American ballet dancer. He was describing the Hollywood movie star and legendary hoofer, who danced with just about every leading lady in twentieth-century American film, and also with brooms, rotating rooms, firecrackers, even his own shadow. He inspired a whole generation of Americans to get out there and dance the night away, with a chain of franchisable dance studios trumpeting the cause. As a brain scientist, watching his seemingly effortless movements, I say he should inspire us again. Unfortunately, he died in 1987, at the ripe old age of eighty-eight.

The reason for my enthusiasm is scientific. You can cover the dance floor with peer-reviewed papers showing the benefits of this regular, ritualized movement that forces social interaction. The scientific benefits are almost too good to be true.

Consider one study, where researchers enrolled healthy older adults, ages sixty to ninety-four, in a six-month dance class, one hour per week. The investigators assessed a broad range of cognitive and motor skills before class commenced, then assessed them again six months later. Non-dancing controls were also measured.

The results were as welcome as free tickets to the Bolshoi. Hand-motor coordination (as measured by a standardized Reaction Time Analysis assay) improved by about 8 percent in six months. That might not sound like much, until you consider that the scores of the controls actually *decreased* during the same period. Suites of cognitive skills were tested, including fluid intelligence, short-term memory, and impulse control. These increased by an impressive 13 percent during the dance class. Posture and balance (measured by using the so-called forced-platform test) increased by about 25 percent in the dancers over their previous scores. And again, the nondancers showed a net decrease. Half a year later, the dancers did not move the same way—or think the same way.

The type of dance didn't seem to matter. Tango, jazz, salsa, folk, various kinds of ballroom dancing: all exerted their whirling wizardry on the brain. Further research has shown that other forms of ritualized movement instruction, such as tai chi and various martial arts, also show benefits in many of these same measures.

One of the most unexpected findings had to do with the number of falls experienced by seniors who took movement classes. During the testing period in one tai chi program, the number of falls fell by 37 percent. Falling is not a trivial issue for the elderly, and for the two reasons they care the most about: head injuries and bank accounts. In the United States, medical expenses from seniors' falls total more than $30 billion a year. In Australia, fall-related injuries among the elderly take nearly 5 percent of the health care budget.

Fred Astaire was obviously on to something.

The human touch

Why does dancing work? The truth is we're not sure. Undoubtedly exercise plays a part. Dancing requires participants not only to learn and memorize synchronized coordinated movements but also to muster up the energy to perform them. There are socialization arguments to consider, too. In most of these studies, a room full of people would be dancing, often as partners, requiring at least a two-drink minimum equivalent of social interactivity.

Finally, there is the idea of face-to-face interactions. And here we have something of a surprise. Depending on the style, dancing allows the opportunity for a certain amount of human touch. That's important for anybody, but it's wildly important for the elderly. The benefits of touch for senior brains—and just about everybody else's brains—have been studied in the laboratories of such notable scientists as Dr. Tiffany Field, director of the Touch Research Institute at the University of Miami. She didn't study dancing. She studied massage, and was among the first to show powerful cognitive and emotional boosts associated with the practice.

Virtually everybody Field has ever tested has shown the benefits of touch, from our oldest citizens in nursing homes to our youngest premature citizens in NICUs (neonatal intensive care units).

Field didn't have to hire a formal masseuse to get the benefit. Even infrequent touching by nonprofessionals, like your friends, helps cement relationships (if the touch is welcome, not exploitive). Fifteen minutes a day will do. That may help explain the invisible devilry of the dance floor, for you often get and give much more than fifteen minutes of touch.

This leads to some practical advice. If you are a younger person, learn how to dance, then keep up the activity clear into your retirement years. If you are already old enough to think about retirement, this recommendation is even stronger. If you already know how to dance, find a place where you can cut a rug regularly. And if you don't know how to dance, take a class, then start your rug cutting.

This helps us settle a digital question, too. As you know, I think social media is a country for old men and women, especially poignant for the mobility impaired. Yet the preferential power of face-to-face communication is clear. Whenever there is a choice to have it, choose it. When at all possible, allow other humans to share the same oxygen as you. Yes, such contact has its pitfalls, but it is what the brain needs in its twilight years. You may feel awkward on a dance floor. You may feel awkward talking instead of typing. Yet for the millions of years we have evolved, we had flesh-and-blood interactions, not server-and-CPU interactions.

Considering the power of socialization on the brain, being with each other is the most natural thing in the world.

SUMMARY

Be a friend to others, and let others be a friend to you

- Keep social groups vibrant and healthy; this actually boosts your cognitive abilities as you age.
- Stress-reducing, high-quality relationships, such as a good marriage, are particularly helpful for longevity.
- Cultivate relationships with younger generations. They help reduce stress, anxiety, and depression.
- Loneliness is the greatest risk factor for depression for the elderly. Excessive loneliness can cause brain damage.
- Dance, dance, dance. Benefits include exercise, social interactivity, and an increase in cognitive abilities.

your happiness

brain rule

Cultivate an attitude of gratitude

Wrinkles should merely indicate where smiles have been.
—Mark Twain

Happiness is nothing more than
good health and a bad memory.
—Albert Schweitzer

A BIRTHDAY CARD RECENTLY caught my eye: "Grumpy Old Man To-Do List."

1. Tell kids to get off MY lawn.
2. Scowl at the neighbor.
3. Write SCATHING letter.
4. Disinherit somebody.
5. Go for a long SLOW drive in the passing lane and keep signal on the whole time.
6. Tell kids to get off my lawn AGAIN!
7. Buy more NO TRESPASSING signs!
8. Tell some punk that in my day we had it tough.
9. Grumble grumpily for a while.

You open the card and it says:

10. Have a Happy Birthday!

As this card suggests, with many all-capped letters, older people have a reputation for being grumpy. Is this reputation warranted?

Seniors *also* have a reputation for being kindly, patient, and wise—words not usually associated with grumpiness. That was certainly my experience with my grandparents. From a research point of view, these questions have serious definitional issues. What does happiness even mean? While researchers don't unanimously agree on definitions, I am going to go with research psychologist Ed Diener, who defines happiness as "subjective well-being." And with legendary researcher Martin Seligman, who defines optimism as knowing that bad things don't last forever, that good will return. One is a condition of the present, and the other is an attitude about the future; both perspectives seem useful. As we'll see, our thirst for optimistic experiences—and our ability to recall them—grow more robust as the years go by.

Onward, mostly upward

Confusion reigned for the longest time about whether people got grumpier or happier or just stayed themselves as they aged. Some studies found that people really fit Beatrix Potter's classic "grumpy gardener Mr. McGregor" stereotype: they got crankier as they got older. Perhaps this was because the seniors studied lived in an environment of unrelenting arthritis, unrelenting funerals, and unrelenting loneliness. Other studies seemed to show the opposite. People became happier and better adjusted as they aged, becoming the type of sage that actor Morgan Freeman often plays in shows like *The Story of God*. Perhaps this was because they lived in a world of increasing wisdom, found a way to avoid more heartaches, and became more socially enriched as they shared their insights. Which is it, folks, Beatrix Potter or *The Story of God*?

Happily, further research provided a clearer picture, and much of it is positive. People really do become happier as they age, but with one important caveat about depression, as I'll explain shortly. They develop more emotional stability, become more agreeable, are more conscientious. The difference is not small. To take just one psychometric measure, people in their sixties score 69 percent higher than people in their twenties on emotional stability assessments. Seniors score even higher on agreeableness tests.

Why the historical discrepancy? It's a classic error. Most of the older studies did not take into account the environmental life experiences of those they examined. This includes controlling for what we now call the usual socioeconomic suspects: wealth, gender, race, mood, education, job stability—even year of birth. Seniors born in the Great Depression, for example, don't have the same happiness profiles (a chart of the years they tended to be most and least happy) as baby boomers, and both have profiles different from millennials'. Whether you have children is a factor, too. Marital satisfaction, which profoundly influences happiness assessments, ebbs and flows depending on the age of the kids you are raising. Marital happiness is highest when the kids are gone, by the way—in the stretch of life between empty nest and retirement. It's lowest when the kids are teens.

When you wade into the deep end of the statistical pool and take some of these factors into account (as was done in one National Institute on Aging study, which looked at several thousand people born between 1885 and 1980), a clear upward trend toward happiness emerges. As one journal put it, "Well-being increases over *everyone's* lifetime" (emphasis mine). Another study taking into account similar variables—this one involving more than fifteen hundred people ages twenty-one to ninety-nine—also found that people aged on a positive note. And if that were the end of the story, we could just whistle a happy tune, pack up our bags, and end this chapter. Turns out not everything about mood improves, and the boost does not last forever—or for everybody. Before we get to that, however, we have to figure out why, for so many, it lasts so long.

What Satchmo says

One of the most relentlessly upbeat songs of the late 1960s and early '70s rock era was performed not by a rock group but by a jazz legend. It was Louis Armstrong's interpretation of "What a Wonderful World":

> *I hear babies crying,*
> *I watch them grow;*
> *they'll learn much more*
> *than I'll ever know.*

Armstrong then marvels at what a wonderful world it is. Some people took exception to this half-full glass of rosy-colored water. With the Cold War in full flower and the Vietnam War in hideous bloom, the world could hardly be called wonderful, right? Armstrong heard about these criticisms, of course, and before graveling it out at a concert one night, he announced this to the audience:

> Some of you young folks been saying to me: "Hey, Pops—what do you mean, what a wonderful world? How about all them wars all over the place, you call them wonderful?" But how about listening to old Pops for a minute? Seems to me it ain't the world that's so bad but what we're doing to it, and all I'm saying is see what a wonderful world it would be if only we'd give it a chance. Love, baby—love. That's the secret.

Remarkable, coming from a man whose greatness endured large helpings of Jim Crow, using bathrooms and drinking fountains marked "Colored Only."

It is Life 101: we'll experience both positive and negative events over time. The same generation that witnessed the My Lai massacre also watched a man land on the moon. As the years roll by, however, our brains don't process positive and negative information in a balanced way. Our desire for (and memory of) optimistic input gets more intense as we age, and we begin to experience life more as a wonderful world.

How do we know? The surprising initial finding of scientists was that older people experienced fewer negative emotions than their younger counterparts. Researchers such as Mara Mather, gerontologist at USC, and Laura Carstensen, director of the Stanford Center on Longevity, decided to investigate. Consistently, they found that older people's brains paid more attention to positive stimuli than negative stimuli. And seniors remembered more details about the optimistic stuff, too.

One experiment involved younger people (average age twenty-four) and older people (average age seventy-three) gazing at happy and sad faces. Which would they pay the most attention to ("attentional bias")? When the youngsters looked at positive faces, they scored 5 out of 25 on the bias scale; they scored 3 out of 25 for negative faces. This meant they were paying modest, fairly balanced attention to each. When seniors looked at the same faces, they scored 15 out of 25 for positive faces and −12 of 25 (yes, *negative* 12) for the not-so-positive faces. Nothing modest or equal there.

Researchers observed similar differences when they assessed that most dissonant note of aging neurons—negative memories. To understand these data, we need briefly to review how memory works. (We'll have a more full-throated treatment in the memory chapter.) The important concept is this: brains don't record life as if on a single reel-to-reel tape deck. Rather, many semi-independent memory subsystems exist—many types of tape decks, if you will—each responsible for recording and retrieving a specific domain of learning. Learning how to ride a bicycle, for example, uses a different neural deck than remembering an episode of *Breaking Bad* or recalling that Tony Bennett sang "Put on a Happy Face." Your ability to recognize something you've seen before (recognition memory) uses yet another memory subsystem.

To test recognition memory, both younger and older populations were shown pictures of "positive images" and "negative images" (such as a person making a happy or a sad face). Younger adults recalled both at roughly equal percentages. Not so for older adults. Their recognition scores were 106 percent higher for positive images than for negative ones.

Researchers have noticed analogous changes for episodic memory (memories for events), short-term memory (now called working memory), and long-term memory (just what it sounds like). The phenomenon even has a name: the Positivity Effect. One reason older people report being happier is that they're increasingly selective about what they pay attention to—and what they remember when they do pay attention.

Why all this optimism in seniors? After all, their joints begin to ache for reasons that become increasingly untreatable, friends begin dying off as if in a war zone, they forget why they went downstairs, and they quit remembering your birthday. Happiness is probably one of the many rewards the brain uses to keep us pro-social. Emphasizing positivity keeps depression away, buffering against suicide. People who are more positive toward us are more likely to lend us a hand in our old age—useful for survival.

There's another pro-social reason seniors are happier. To explain it, I'll turn to an Industrial Age Brit who would not be caught dead putting on a happy face. We're going to talk about that quintessential grumpy old man, Ebenezer Scrooge.

Lessons from London

The most unsettling aspect of Dickens's *A Christmas Carol* to me is that some of its nineteenth-century pages seem lifted straight out of a twenty-first-century geroscience textbook. As proof, I offer you a few degrees of Ebenezer Scrooge's famous narrative arc. He starts as a miser about Christmas, as you know, and doesn't change to Santa Claus until he finally confronts his death. What helps turn him away from the dark side isn't a grave marker, however. (Presumably death is as much a concern to innocent Tiny Tim as to greedy moneylenders.) The change agents come gradually, from the types of things Scrooge observes as the ghosts force-feed him his biography. When Scrooge is young, his mind is on his newly minted career, the knowledge-based world of industrial-age banking—and his increasingly self-centered success. But when he is old and the Spirits have held sway, his priorities have been turned upside down (or, rather, right side up). He exchanges the cold, knowledge-soaked world of accounts receivable for the warm, emotion-soaked earthiness of human relationships.

Here, data and Dickens meet, for this exchange—minus the ghosts—is exactly what happens to our brains with age. We, too, shift from paying off college debt and other financial priorities to playing with grandkids. This, on average, makes us happier. The delightful metamorphosis stems from both nurture and nature, each of which deserve a hearing.

In your youth, your brain fools you into believing you'll live a long time, if not forever. This is an attitude positively shackled with social consequences—ranging from whether you commit to retirement savings or sign up for health care. (Insurance companies often call people in this age group "the immortals.") You are also at the starting gate of your career, and so you see knowledge-based pursuits as your top priorities for future achievements. Ditto for your relational successes. Anyone who's been married, had children, or experienced both understands how much extra knowledge you need to be successful.

All that changes with age. You now have a few more miles on your biological tires and greater knowledge about how the world works. You hardly need three ghosts to realize you were wrong and won't live forever. I remember first discovering this when I wrote down the number of books I wanted to read before kicking the bucket. I calculated the time it would take to read them and realized I would need to live more than 180 years to finish. And that's if I didn't do anything else but read books. While that is surely a vision of heaven for me, I have, unfortunately, other things to do. Aging forced me to prioritize. And since I knew I wanted to spend more time with my family than with Dickens, or any other author, I could sense something warmly relational shifting beneath my behavioral feet.

This shifting is consistent with the research literature. When you truly realize you have a sell-by date, like the post-ghost Scrooge, you begin to prize relationships over just about everything else. And any time you prioritize the socioemotional components of life, you become happier—the entire point of the friendships chapter. This shift is so common, and is backed by so much empirical support, it's been christened with its own tediously academic title: socioemotional selectivity theory.

At the same time scientists puzzled over the weight of these behavioral data, others began ruminating over their potential neurological origins. They came up with their own, arguably more disturbing, name for their findings: FADE, short for frontal-amygdalar age-related differences in emotion.

We've already discussed one of those differences: the more social relationships you acquire, the bigger the amygdala becomes. Other differences also accrue with age. Aging brains activate the appropriate emotions with greater strength—just as those emotions are changing the way we react to the world. It's quite possible that the neurological effects termed FADE directly affect what we think is important. A lifetime of Christmas geese results.

Roller-coaster grandpa

Older people are supposed to be allergic to risk. Don't tell that to Gary Coleman, a retired pastor from Ohio, however.

Looking something like the actor Sean Penn, if Sean Penn were seventy-four, Reverend Coleman is a roller-coaster fanatic. In 2015 he took his twelve thousandth spin on Ohio's legendary Diamondback roller coaster. "I thought it was the best coaster I've ridden on my whole life," he exclaimed in an interview. "At my age, it's great!" He knows of what he speaks. He's been riding roller coasters obsessively since childhood.

Researchers have found two interesting patterns concerning how risk-related behavior changes as people age, and they're definitely related to happiness, just like the good reverend's roller-coaster experience. One is called the "certainty effect," the other "prevention motivation."

The certainty research was initially hobbled by uncertainty. That's because young and old are willing to take risks at roughly equal rates, and with roughly equal enthusiasms. Knowing that equality doesn't always mean similarity, however, the researchers put their heads down and charged straight at their spreadsheets. That's when they smashed into something solid. The kinds of risks the generations take are as different as a noisy casino is from a cozy teahouse.

If you are of a certain age and find yourself feeling risk averse these days, you're not alone. When given a choice between larger potential gains with substantial risk, or smaller potential gains with smaller risk, seniors almost always choose the smaller gain. In fact, risk aversion is *always* largest if there's a threat of losing a potential reward, however small that reward might seem. Why? Seniors prefer the higher probability of experiencing a positive emotion. Like with operating the penny slots, reward size doesn't matter to seniors, as long as they can play. This finding is so common, researchers call it the certainty effect.

Contrast this easy satisfaction with our younger selves. In youth, the altitude of happiness we experience is consistently stratospheric, and we want more. We crave a dance, a rave, loud music, louder friends. After all, we might find a lifelong mate in these vigorous activities or, later, potential connections for work advancement. It's a risky way to play at life, given the stakes, and aggressively self-serving. But it's also understandable. In our youth, our emphasis is on the future, not the past—perhaps because we haven't made one yet. Which is why staying home and watching reruns of *I Love Lucy* is not exactly our idea of a good time. The researchers who quantified this preference named it "promotion motivation."

Reaping the consequences of promotion motivation, we then swear fealty to the iron thrones of mortgages and parenting and saving for retirement. We turn into efficiency experts, seeking ways to preserve success and prevent failures as both begin accumulating in larger numbers. We become concerned with what we can keep as much as what we can create. Eventually, the vaguely disturbing illusion that we'll live forever vanishes. Heading into retirement, we try to protect what we've worked so hard to gain, shifting from promotion motivation to prevention motivation.

It's a good name, goaded by the ultimate irony of life: death. Now we see ourselves in terms of preservation *because* time is short. Present happiness becomes more important than future reward. With creaking joints, friends dying, and loved ones moving away, a night with Lucille Ball might just be the ticket.

In a nutshell, that's the relationship these shifting feelings have with risk taking. We become eager to shun potential hazards and embrace smaller rewards simply because we may not have many rewards left to enjoy. After the twelve thousandth time on a roller coaster, you realize that it's not going to harm you and there's still plenty of joy to wring from the experience. What, then, is the harm in making it 12,001?

Smell a rat?

As I mentioned, I have not told you the whole story about seniors and happiness, and there's a reason for my cowardice: it's not all razzleberry dressing and theme-park rides. Here's a true example of how sad this not-good-news can be.

A widowed seventy-four-year-old physician from Southern California found the loneliness excruciating; he eventually signed up for a dating website. The good doctor quickly found an anesthetic in the form of a forty-year-old British divorcée. She was broke, with a daughter in college. In a few weeks, they were cyber-buddies, and weeks after that, long-distance lovers, the digital equivalent of a winter/summer relationship. You probably already smell a rat, and we might wish the doctor had, too.

The woman contacted him in a panic one day. Her daughter had been killed in a car wreck. She didn't have the money to pay for the funeral, nor her daughter's student loans, and could he wire her $45,000 to cover the costs because she just didn't know where to turn? He sent her the funds—which, of course, opened up a spigot of other requests. A fortnight later it was $10,000 for a new roof, $75,000 for a new Mercedes (yes, *Mercedes*), and finally, a first-class ticket from London, so she could meet the love of her life at LAX and thank him in person. Sadly, he granted every financial request. The woman never showed up at the airport, though he had readied a limousine, Cristal champagne, flowers, and a room at the Four Seasons. The doctor never heard from her again.

This stuff happens to the elderly all the time. Figures are hard to come by, but MetLife estimates the elderly get bilked out of nearly $3 billion every year. Men and women are equally vulnerable, and successful Beverly Hills family doctors are not immune. It's proof that older people should be less concerned about running out of life than running out of money.

There's an obvious reason the elderly become a target: solo seniors sometimes have obese bank accounts. The less obvious reason has to do with the dark side of focusing on the positive all the time. As you age, you also become more trusting, or better to say, more gullible. We even think we know why.

There's an area of the brain called the insula, a slightly hidden clot of neurons just above your ears. You can think of it as an "ability to know when you're being taken for a ride" detector. Like so many brain regions, the insula has other subfunctions, ranging from assessing risks to reacting to betrayals to feeling disgusted. It even helps forecast whether a given action will be safe. As you age, the anterior insula (front area, nearest your eyes) becomes less reactive to potentially untrustworthy, even threatening, situations. Scientists can show the effects of this decline in many ways, including the capacity to detect untrustworthiness in people's faces. Or in fake British lovers.

This vulnerability is part of a general collapse of an extremely important behavior: the ability to know when you're making a mistake about something—especially if rewards are involved. It's part of a general suite of behaviors called reward prediction, which is the ability to forecast when a reward is most likely to happen (or not). Reward prediction abilities decline more than 20 percent with age, which means reward prediction errors increase. (A reward prediction error is where you anticipate that a reward will occur, based on prior experience, but it doesn't, so you are wrong.) And just to complete the circle, the aging brain gets worse not only at predicting rewards but also at assessing risks.

There's more bad news. As if a flickering insula wasn't enough, there is an area in the brain I sometimes call the AC/DC network (the "Highway to Hell" circuit) that changes as we age. The Highway to Hell is a ridiculously powerful series of linked circuits deeply embedded in your brain, close to the insula. The circuits are responsible for many things, including nearly all addictive behaviors. Hence the name. They are also involved in reward prediction errors, mediating what we call "probabilistic learning," a skill at which you get increasingly bad with old age. Researchers believe these two regions, the insula and the Highway to Hell, are the reasons why older people become more gullible. And why people who love us need to take special precautions if they take care of us. An aging insula and accompanying circuitry are as dangerous as a broke lover.

A darker shade of gray

I still remember the first time I heard these lyrics over the car radio: "Ooh, what a lucky man he was." I got goose bumps. I was amazed as the song ended with one of the strangest keyboard sound clusters I'd then encountered. I didn't normally listen to rock in those days—still don't (I prefer Stravinsky to the Stones), but I wanted to know more about this group. It was a trio, sporting a name more like a law firm than one of the great prog-rock groups of the '70s: Emerson, Lake & Palmer. When I discovered they also did electronic covers of classical pieces, it was love at first toccata. I was especially enamored of the virtuosity of the group's legendary keyboardist, Keith Emerson. Thus it was with sadness that I read about Emerson's suicide, in 2016, at age seventy-one. Though he kept the dogs of depression at bay for years, his resistance ended when he developed career-threatening nerve damage to his fingers. Gun in hand, he became not such a lucky man after all.

Depression and suicide go hand in hand, as Emerson's life illustrates. Depression and old age go hand in hand, too, which his life also illustrates, and this represents the deepest shade in the dark side of our chapter on happiness. It also seems to contradict virtually everything I've been discussing so far. I obviously have some explaining to do. And with help from two quotes in the research literature, I intend to do just that.

First, we need a quick definition of depression. That's important because people often confuse depression with normal sadness. In fact, seniors in the grips of depression often *don't* feel particularly sad. Instead, they become increasingly unfocused and demonstrably more irritable and restless, and they experience a steady erosion in things they used to find pleasurable. We also need to take into account the fact that triggers for depression—health failures, deaths of loved ones, unremitting pain—are routine events for the elderly.

Older literature about senior depression, such as our first quote (from the surgeon general of the United States, circa 1999), says things like: "Depression is not a normal part of aging . . . serious depression is not 'normal' and should be treated." True? Though the appeal for treatment is spot-on, later research showed the rest of that quote is true only if you don't look too closely. If you do look closely, you run right into our second quote (from researcher Ke-Xiang Zhao, at Chongqing Medical University in China), which takes issue with the idea that depression isn't typical: "Older age appears to be an important risk factor for depression in the general elderly population (aged below 80 years)."

Reconciling these seemingly different perspectives, it turns out, depends on how often you had to visit the hospital. For moderately healthy seniors, depression isn't typical. For seniors whose health is impaired, it's a different story. (And it's a good thing that researchers made the distinction, because if they lumped everyone together, they could be fooled into thinking they're looking at "natural erosion" rather than "unnatural disease progression.")

Here's what we know now: the more health challenges seniors encounter, the greater their depression risk becomes. The type of disability is the major contributor, with chronic disease taking pole position. One of the biggest contributors to depression is hearing loss. Another biggie is vision loss. Others are the various cancers, chronic lung diseases, strokes, and cardiac diseases. Unknown are the effects of diabetes and hypertension.

If seniors live in community settings, depression chimes in at a modest 8 percent to 15 percent. Hospitalize them because of some physical ailment, or simply put seniors into assisted living, and the prevalence soars to 40 percent. That's a big deal. Depression is now projected to be the leading cause of disease burden in the elderly by 2020. The bottom line is that happiness increases in older populations as long as seniors remain healthy. But since health naturally ebbs in aging populations, the rate of depression rises.

Is there something we can do? Though the answer is yes, we must revisit some brain biology to understand our options, examining one of the happiest biochemicals on earth. Would that Keith Emerson could have become better acquainted with it.

Dopamine's decline

"*That's* the problem," my dad chuckled one cold winter morning in 1966, holding up a small, jewel-like bauble for me to inspect. It looked like the threaded end of a decapitated Christmas light. "If we replace the old guy with this one, the kitchen's gonna work good as new."

Earlier that morning, my ten-year-old self had marched into his bedroom, horribly alarmed that I had broken the entire kitchen. I had plugged in a portable space heater near the fridge, then heard a loud pop. The kitchen immediately stopped working. No lights, no refrigerator, no stove, no electric can opener.

"All you did was blow a fuse, Son," my dad said, fingering his glittering electrical ornament, a spare (now vintage) fifteen-amp household fuse. I was amazed. How could such extensive culinary destruction—from refrigerators to ovens—result from something so small, so singular? I got my first lesson in how electrical circuits worked in houses. Dad unscrewed the old fuse and put in the new one; sure enough, the kitchen roared back to life.

This electrical nostalgia illustrates something useful about brain wiring and its activating circuitry. I've mentioned many behavioral changes in this chapter: decision making, award seeking, risk taking, selective memory, depression. These behaviors might seem as functionally disconnected as a can opener from a freezer. But they aren't disconnected at all. Scientists believe the biological basis for most of these changes comes from the failure—just like in that kitchen—of a single circuit.

This circuit isn't made of wires responding to electricity, of course; it's made of neurons responding to a neurotransmitter. The neurotransmitter is a famous molecule I'll bet you've heard of before: dopamine. The circuits over which dopamine exerts its powers are called dopaminergic pathways. The brain has about eight of these pleasure-coaxing pathways.

One of the first impressions you'd get if you ever bumped into a molecule of dopamine is how ridiculously small it is. It's synthesized by redecorating an amino acid called tyrosine. Remember amino acids from high school biology? They're the natural building blocks of proteins. To make a protein, long strings of amino acids—sometimes hundreds—are strung together like cars in a train. Dopamine is the size of just one of those train cars.

You may also be familiar with tyrosine because of your diet. Most of you eat it every day. Egg whites have a lot of tyrosine. So do soybeans. And seaweed. Don't be fooled by its size or pedestrian origins, however. Dopamine packs a serious wallop. Make too little of it and you might get Parkinson's disease. Make too much of it and you might get schizophrenia. When you synthesize just the right amounts, dopamine mediates your ability to reward yourself with pleasure, your ability to hold a pen without shaking, and your ability to make decisions. Every one of the behaviors mentioned in this chapter at some level involves dopamine. Impressive skill set for a clump of seaweed.

How does this polymath of a molecule do it? Dopamine mediates its activities by binding to a family of receptors built for it. These receptors are found only on certain neurons in the brain. Cells lucky enough to sport the receptors are activated to perform certain functions when dopamine binds to them. Think of it as the ignition system inside your standard Honda. Insert the key into the lock, and the car springs to life. Insert the dopamine into its neuron-bound receptor, and the neuron springs to life. Put many of those neurons in a row, and you have an activatable circuit. Put eight or so of those circuits together, stuff them deep into the center of the brain, and you have the dopaminergic system.

Given the brain's Shanghai-esque overpopulation of cells, the dopaminergic system involves remarkably few neurons. Only certain regions contain dopamine receptors, which means only certain regions of the brain are sensitive to dopamine. One prominent area is the "Highway to Hell" circuit I mentioned. This highway consists of two small dopamine-sensitive brain regions (the ventral tegmental area and the nucleus accumbens) connected by dopamine-sensitive circuits. Over-driving this system—and thus dysregulating it—is responsible for most of the chemical addictions that regularly devour human beings.

Dopamine, it turns out, is a really big deal. And we are about to find out just how big a deal it is for seniors. One of the hallmarks of aging is that the dopamine system, after a while, begins to fade away.

The mouse that didn't roar

Some experiments are tough to digest, like an overcooked steak, and this is one of them. You can genetically manipulate mice in such a fashion that they can't make dopamine by themselves. When you do that, you give them a death sentence. The reason is startling. *The animals starve to death*. Even if you put their favorite foods in front of them, the rodent equivalent of chocolate cake, they will sit there beside the food, blinking at it, doing nothing to intervene as death slowly envelops them. Same for baby mice. Without intervention, dopamine-deficient pups won't suckle frequently enough to sustain their little lives. They still have the behaviors necessary to look for the food and eat. They just aren't willing to eat. Intervene by administering dopamine artificially, and everybody starts to eat normally. The point? Life without dopamine can be very difficult to sustain. Life with dopamine is, to understate the obvious, the preferred option.

The reason I bring up this experiment concerns one of the most solid biological findings that exists in the gerosciences: as humans age, the dopaminergic system begins functionally to decline. In humans, this decline has consequences much more complicated than simply changing our pleasure in eating. Since the human brain has a cortex the size of a baby blanket, and lab rodents have one the size of a postage stamp, such differences make sense.

The erosion in humans has three parts. First, the manufacture of dopamine slows down in specific regions of the brain. It's an uneven assault. Midbrain loss is smaller, whereas loss in the forehead-dwelling dorsolateral prefrontal cortex is almost threefold greater. The effects are especially noticeable after age sixty-five. Second, dopamine receptors begin to disappear. One important receptor, dubbed D2, declines 6–7 percent with each decade of life, beginning around age twenty! Third, dopaminergic neural circuits begin flickering off, mostly because of cell death. One commonly hard-hit region is the substantia nigra, a piece of neural real estate deeply involved in motor function. Parkinson's disease can result, which explains why one of the greatest risk factors for getting it is simply growing old.

These three categories of losses may explain virtually every behavior discussed in this chapter. Certain types of depression occur because of a loss of dopaminergic activity, for example. So common is the experience, it's been given its own name—DDD, for dopamine deficient depression.

We also know that dopamine is involved in mediating command decisions—particularly reward prediction—which, as you recall, is a skill that diminishes with age. Dopamine mediates the willingness to take risks, which also declines. Dopamine is even associated with our psychological motivations. Given that age transports us from aggressive promotion motivation to cautious prevention motivation, we may be observing the alteration of a singular suite of risk-associated behaviors.

Even the positivity effect (and its darker twin, gullibility) may be explained by dopamine loss. We know that attentional networks, which allow us to preferentially select one set of stimuli over another, are profoundly influenced by dopaminergic activity. Indeed, most of the major players in those networks use dopamine to direct the focus of our brains. That includes the insula (coincidentally also involved in gullibility), which in youth is studded with dopamine receptors like cloves on a ham. A dysfunctional insula, by the way, is also associated with depression.

What about seniors who report being happier as they get older? Does dopamine dysregulation play a role here, too? The real answer is we don't know. As we've seen in this chapter, the happiness data are nuanced, especially when other factors are considered (like diseases and depressions). Since these studies were done primarily with healthy seniors, "healthy" may also include intact dopamine pathways. In which case, scientists were studying only a subset of the population.

Or not. As we'll see in the memory chapter, the brain is surprisingly good at conjuring up compensatory behaviors for cognitive functions it knows are eroding. The happiness data may represent the determined effort of a brain, faced with inexorable dopamine decline, refusing to go down without a fight. Or a smile. Many seniors I know still light up in the presence of chocolate cake and start looking for a fork. I'm one of them.

Awakenings

While scientists in various corners of the research world are actively investigating these processes, others have skipped ahead of the biology and gone right to the clinic. They are interested in determining what, if anything, could be done practically for patients now. If dopamine loss is so deeply associated with behavioral decline, they asked, could that decline be arrested by artificially resupplying the molecule? Research suggests there might be something to this idea.

One of the most surprising examples of this practical approach originates from a 1973 book called *Awakenings*, a true story written by famed neurologist Oliver Sacks, made into a movie years later.

The book wasn't about patients suffering the consequences of aging. It was about patients suffering the consequences of infection (encephalitis). The disease left most of the patients catatonic, wheelchair bound, seemingly alive in name only. When one of these closed-for-business people (played in the movie by Robert De Niro) was administered a synthetic form of dopamine, it was like giving him a syringe filled with the Fountain of Youth. He suddenly awoke from his catatonia. He started smiling, walking, talking, wanting to fall in love—Sleeping Beauty responding to a kiss from Prince Dopamine.

This synthetic dopamine, biochemical royalty in the world of neuroscience, is called L-DOPA. (You can't use real dopamine because it strangely refuses to jump into the brain.) L-DOPA has triggered at least two Nobel prizes, mainly for treating Parkinson's disease. Studies have also shown positive effects on cognitive processes not associated with disease states but simply with typical aging.

Consider reward prediction, which withers with age. You can alter its fall from grace by taking L-DOPA, literally improving a complex cognitive process on the back of a simple synthetic. The effect isn't small. The treated seniors' laboratory performance becomes indistinguishable from that of younger, untreated controls.

L-DOPA increases your preference for looking on the Doris Day side of life, too. It elevates something called optimism bias, prejudices about which seniors know a great deal. But this experiment was not done with seniors. It was done with a younger generation, known more for loving snark than *Singin' in the Rain*. It caused the author of the experiment to declare: "This study does show that optimism may be influenced by dopamine levels even in healthy people. And that's a pretty glass-half-full kind of study."

That's especially good news for seniors. Optimism is not just emotional insulation against the freezing wastes of mortality. We now know that elders who have positive, even optimistic, attitudes toward their own aging live longer than those who don't.

What do I mean by optimistic aging? A twenty-five-year-old who forgets somebody's name seldom considers it a harbinger of Alzheimer's disease. But if you're older and your memory transmission slips a gear, you might very well worry about Alzheimer's. You may become stressed, even depressed. As other roadside attractions of age come into view—from hearing loss to aching joints—your attitude may turn increasingly pessimistic. The data say: don't go there. Seniors who take it in stride, convincing themselves the glass is still half-full, live a healthy 7.5 years longer than seniors who don't. Optimism exerts a measurable effect on their brain. The volume of their hippocampus doesn't shrink nearly as much as the glass-half-empty crowd's does. That's an important finding. The hippocampus, a sea-horse structure located just behind your ears, is involved in a wide variety of cognitive functions, including memory. My guess is that dopamine levels are affected, too. These seniors avoid the trap of what would otherwise turn out to be a self-fulfilling prophecy.

And you don't need a drug to practice optimism.

That leads to an important question: Should you rely on drugs in an effort to set this attitudinal shift in concrete? The movie *Awakenings* may be instructive here, too, as it's based on something that really happened. The effects of L-DOPA turned out to be temporary. Robert De Niro's character eventually returned to his catatonic state, as did all his colleagues. The movie ends with one of the saddest dances in film. Though it has done great good, L-DOPA comes, as all drugs do, with some important side effects, including hallucinations and psychoses—and in the case of encephalitis, term limits.

Are there ways to maintain optimism (and possibly elevated dopamine) that do not require drugs? Something that yields a more permanent, more side effect free future? The answer, happily, is yes. It all revolves around the hidden part of what scientists mean when they say, "Don't go there."

Oprah Winfrey had an unpleasant childhood, to put it mildly. She still remembered those difficult roots when she became famous, lending her rags-to-riches story authenticity. She once said: "Though I am grateful for the blessings of wealth, it hasn't changed who I am. My feet are still on the ground. I'm just wearing better shoes." Consistent with this attitude, Winfrey started jotting down all those blessings, a journaling habit she sustained for a decade. There are scientific reasons why it was good that she did. Winfrey probably knew this: her emphasis on gratitude ran smack into some solid cognitive neuroscience, enrobed in a body of thought called positive psychology. The research I'm describing comes from its father, Martin Seligman, who used to study trauma and depression.

Observing the megawatt power of gratitude as a practicing psychotherapist, Seligman developed—and then scientifically tested—

exercises centering on the ideas of thankfulness and appreciation. Here are two famous three-steppers worth trying:

The gratitude visit

1. Find someone living who has meant a great deal to you.
2. Write that person a three-hundred-word letter. Describe concretely what he or she did to make you want to pen the letter, and explain how that still influences your life.
3. Go visit the person, letter in hand. Read it aloud (without interruption), then discuss.

The effects, Seligman found, are as quick as laughter. A "happiness psychometric inventory" (yes, those exist) found a noticeable boost in the writer's happiness a week after the visit. The effects lingered even one month later.

"What went well" (or "Three good things")

1. Recall three positive things that happened to you today.
2. Write them down. They can be smaller ("my husband brought me coffee") or larger ("my nephew got into the college he wanted").
3. Beside each positive event, describe *why* the good thing happened. "My husband loves me" might be written beside the coffee comment. "My nephew worked his butt off at school" might go next to the college comment.

Do this every night for a week.

This exercise can be quite powerful. It not only boosts happiness scores but also successfully treats depression. The elevation takes longer to observe (about a month), but it also lasts longer. Though the experimental exercise lasted only a week, improvements were still measurable six months later. If these gratitude behaviors become habit, so do their long-term benefits. Here's how Dirk Kummerle of the Massachusetts School for Professional Psychology couches the findings: "[The] gratitude visit and three good things were not only able to reduce depressive symptoms (compared to subjects with no intervention) but also provide lifelong tools to combat negative thoughts and cultivate well-being."

These exercises provide connective tissue to a powerful research goal: understanding what makes people authentically happy. Seligman has codified the science into what he calls "well-being theory." It is composed of five contributing behaviors, summarized in the acronym PERMA. These represent an actual recipe, a to-do list for people of any age interested in authentic happiness—but perhaps especially useful for those whose dopamine systems are currently being gutted. I provide only a summary here; I encourage you to read about the research directly in Seligman's book *Flourish*.

P: Positive emotion

To be happy, you must regularly experience positive emotions. Generate a list of the things that bring you true pleasure, then marinate yourself in them, allowing the items on the list to become a regular part of your life.

E: Engagement

Consistently engage in activities so meaningful you actually stop checking your cell phone when you do them. Losing yourself in a hobby can be like that. So can good movies, books, sports—even a dance class.

R: Relationships

As long as the relationships are positive, insert the entirety of the chapter on friendship into this recommendation.

M: Meaning

Identify and pursue a purpose that gives your life meaning. For most people, that requires solidly connecting their actions to a purpose larger than themselves. Religious practice and charitable work are examples.

A: Accomplishment

Set specific goals for yourself, especially if that requires you to achieve mastery in something over which you currently have no mastery at all. This could be physical, like training for a marathon, or intellectual, like learning to speak French.

You can see a lot of Winfrey's life in the midst of these research findings, which is why I bring her up. Now in her seventh decade, she is doing a lot more than just wearing better shoes.

Research shows you should, too.

SUMMARY

Cultivate an attitude of gratitude

- Older people tend to score higher than younger people on clinical tests aimed at measuring happiness.
- The positivity effect is the phenomenon in which older people selectively pay much more attention to positive occurrences in their surroundings. They tend to remember these positive occurrences much more than negative ones.
- As you age and realize your own mortality, you tend to prize relationships above anything else. Prioritizing these relationships makes you happier. This phenomenon is called socioemotional selectivity theory.
- The risk of depression increases in seniors who face more health challenges—hearing loss, for example—than in healthy seniors.
- Optimism about one's own aging exerts measurable, positive effects on the brain.

THINKING BRAIN

your stress

brain rule

Mindfulness not only soothes but improves

Some people have told me that I'm grumpy; it's not something that I'm aware of. It's not like I walk around poking children in the eye … not very small ones, anyway.
—Irish comedian Dylan Moran

Worrying is like a rocking chair. It gives you something to do, but it doesn't get you anywhere.
—Anonymous

IF THERE WERE A contest for Most Interesting Man in the World, my grandfather easily could have won. He journeyed to North America as a ship stowaway, complete with an aristocratic Spanish accent, arriving penniless. His mind was well funded, though: rich with humor, radiant as the sunny Meseta Central, blessed with the ability to pick up any language (I lost track at eight). These attributes helped him secure a place in the food industry, working his way to becoming a sous chef at a Detroit country club. He opened a chain of bakeries, raised his family, and died at 101. The last time my wife and I saw him alive—a 100-year-old still in his own home—he showed off his culinary skill. Cheerfully donning his old apron and whistling away, he made six apple pies—at once! Not only was he the most interesting man in the world to me, but he was probably the happiest.

Which is interesting. One might assume that older people would report being quite bothered by life and its attendant changes, more anxious about health and memory and relationship failings, more stressed in general. That is the exact opposite of what researchers find. Older people report being *less* stressed than their younger counterparts. About 38 percent of all young adults in 2016 (so-called millennials, ages eighteen through thirty-four) report being more stressed than they were the year before. That figure drops to 25 percent with the baby boomers, people born between 1945 and 1960. That number shrinks to 18 percent in the so-called Greatest Generation (parents of the boomers), the lowest figure for any group. And they aren't just less stressed. As we saw in the last chapter, older people report being happier. They describe having a greater satisfaction with life, and except for the "oldest old," have lower rates of depression and anxiety.

How could this be? With age, your stress hormones are dysregulating with the fury of a 1930s furnace. Stress is supposed to be like oxygen to the rusting hull of your aging brain. Yet seniors just don't seem to feel it. To understand why, we'll need to explore more deeply the biochemistry behind stress responses, weird-sounding brain regions like the hippocampus and entorhinal cortex, mid-abdomen organs like our adrenal-capped kidneys, and thermostats.

Actually, we're mostly going to talk about thermostats.

Running from the grizzly

Stress responses have one delightful job description: to keep you alive long enough to have sex. Your body has organized all kinds of hormones and cells and neurons into complex, interlocking sets of biochemical feedback systems in pursuit of this long-term Darwinian goal.

Though human stress responses are complicated, there is something simple you can say about them: when you're stressed, your body dumps a ton of hormones into your bloodstream. Epinephrine and norepinephrine (or, if you are from the United Kingdom, adrenalin and noradrenalin) are often the first responders. Wielding immense physiological power, these catecholamine twins stimulate your cardiovascular physiology, increasing your heart rate, altering your blood pressure, and overstuffing your muscles with oxygen. They prepare your body to run away from Mama Grizzly.

This takes a lot of energy, of course, so your body recruits another first responder, the steroidal hormone cortisol, to help control the response. Cortisol is secreted by the adrenal glands, those pyramid-shaped tissues lying atop your kidneys. The elevation of these hormones in your body signal that you are in the grips of a fight-or-flight response, though to be perfectly blunt, it's mostly about flight. Even against a juvenile hyena, we were (and are) too physically weak to put up much of a battle, so we did a lot of running, making us the Pleistocene era's biggest chickens.

Cortisol has an important brain region in its gunsights: the hippocampus. This sea-horse-shaped brain region is famous for being involved in learning. It has custodial rights over the formation of certain memories, such as that bears are real threats. But it's also involved in keeping your stress responses from wearing out their welcome once Mama Bear waddles off to eat berries and not you. Specifically, the hippocampus is involved in ascertaining the first possible moment when it can turn off energy-burning cortisol secretion.

This is a classic negative feedback loop. Proteins called cortisol receptors, which stud the hippocampus like raisins in cinnamon bread, are the mediators. When cortisol is released into the bloodstream, some molecules rush up to the hippocampus and bind to those cortisol receptors, like a key to a lock. The hippocampus is now alerted to the threat situation and is ready with a wide variety of responses.

One of its most important responses is turning off the cortisol spigot, shutting down adrenal activity when the threat is removed. Like spoiled rock stars in a hotel room, stress hormones actually start damaging their host if they overstay their visit. That includes brain damage, by the way. Small wonder that one of the first questions the hippocampus asks when cortisol binds to its receptor is an unfriendly one: "When can I make you go away?"

If the hippocampus ever failed at this job, your cortisol levels would remain abnormally high long after there was no reason to keep them up. That, unfortunately, is exactly what begins to happen to cortisol levels when you age. The hippocampus loses the ability to turn off the hormone.

And that has all kinds of consequences—which is where a working knowledge of thermostats comes in.

Cranking it up

Because I live in Seattle, I am used to lots of moist coolness, even in our warmest month (August). This is the opposite of Houston, where some of my relatives live, which has lots of moist hotness, especially vicious in August. So you can imagine my stress when, at a summer speaking engagement in the Houston area, I found my hotel room's thermostat busted. Or should I say, its sensors were busted. It acted as if an arctic air mass had permanently parked itself in my room, because it kept turning off the air conditioner, trying to heat the room. It was as hot as a freshly baked potato.

As you know, thermostats aren't supposed to work that way. You set the desired temperature, then let the sensors work their magic. If it's too hot, the sensors automatically tell the AC to kick in. If it's too cold, the sensors let the heater take over. This feedback system usually involves tiny strips of metal and the element mercury—and, in my case, a repairman. The hotel immediately called a technician, who fixed the thermostat, and the arctic air soon returned. The device behaved itself for the rest of my visit.

Minus the metal strips and mercury, your stress system has very similar feedback behavior. It even has a set point, though it's more dynamic than my hotel-room thermostat's. Cortisol is normally high when you wake up—anticipating a breakfast filled with predators, perhaps?—and then, if everything is calm, it faithfully depletes throughout the day. It's not a trivial change. On a calm day, there is an 85 percent decrease from morning until evening.

This dynamic system is built to handle only one particular type of stress: a short one. From an evolutionary perspective, that makes sense. The grizzly bear either ate you or you ran away, but it was all over in minutes. It's a finely tuned response, but it's finely tuned only in short bursts.

The problem with modern society is that you can be caught in stressful situations that last for years—say, a bad marriage or a bad job—the physiological equivalent of the grizzly bear moving in with you. I mentioned brain damage. Indeed, exposure to unrelenting long-term stress can lead to major depression and anxiety disorders, which are true collapses of multiple brain systems.

We can graph this idea in the shape of an inverted U. At first, stress responses elevate both physical and mental functioning, the left-hand side of the graph climbing upward, reaching peak performance as long as the stressor doesn't hang around too long. If stress overstays its welcome, the optimization turns into damage and you begin to slide down the ugly right-hand slope of the curve. Even properly functioning stress responses, to normal short bursts of stress, become dysregulated.

There's another way stress hangs around too long: you're outliving a system that was never built to handle life past thirty. Stress dysregulation ends up being a normal part of the aging process—one that is measurable. There are three manifestations.

The first concerns rhythm. Somewhere around age forty, baseline cortisol levels begin to rise. They stop following that lovely morning-high/evening-low rhythm and instead start sloping upward as if skiing uphill. Your body begins to experience the type of damage that occurs whenever stress hormones are elevated. We'll have more to say about that damage in a minute.

The second manifestation is that you don't respond as rapidly—or as vigorously—to the presence of threats. Take your cardiovascular system's reactions to the epinephrine twins. From heart rate to blood pressure, all respond with much less vigor to the "all hands on deck" alerts as you age. You still make as much of the hormones as ever. You just can't respond the way you used to. To make matters worse, once the alert has been sounded and your body begins to obey, the system takes longer to rev up the engine.

Finally, you don't calm down as readily once you've finished reacting. With age, stress hormones have a harder time returning to baseline after a threat. It's as if the aging body says, "Now that you've spent all this effort getting your stress responses elevated, I'll be darned if I'll let them return to earth so soon!"

Do these sound like thermostat issues to you, as if elderly stress responses were acting like my recalcitrant hotel room AC? To explain why, I'll take a scene from one of my family's favorite holiday movies, *A Christmas Story*, which also stars a disobedient temperature-control system.

A dysfunctional damper

The scene opens with Old Man Parker roaring, *"Aha, aha, it's a clinkerrr!"* He's watching black fumes pour out of a basement grate and into his 1930s living room. *"That blasted, stupid furnace! Dadgummit!"* He marches down the stairs to do battle with an obviously rebellious heating system. *"For cripe's sake, open up that damper, will you?!"* his disembodied voice yells from the bowels of urban Hades. *"Who the hell turned it all the way down?! AGAIN?!"*

As you probably know, a damper is simply a flap in the flue of a chimney. Open it, and the smoke from your roaring furnace gets sucked outside. Shut it tight (when the furnace is off), and the damper prevents cold air from coming into the house. Toggle it back and forth, and you control the amount of oxygen available to the fuel source, a crude human-powered thermostat. It's not working in the movie, which is the source of Old Man Parker's increasingly colorful vocabulary. He eventually fixes it, and his profanity is the main price the family has to pay for thermal comfort. The voiceover cheerfully intones, "In the heat of battle, my father wove a tapestry of obscenity that, as far as we know, is still hanging in space over Lake Michigan." Funny scene. Illustrative, too. I'll use it to discuss not only the stress behind the old man's behavior but how his unreliable human-based thermostat explains what happens as he ages. First the bad news, then—I promise—some good news.

The bad news is that when hormones like cortisol remain in your bloodstream, it's like black smoke pouring into your house. Everything is a potential target for damage. Research from many labs shows a single disturbing pattern: the diseases that excess cortisol causes in humans of any age are the same diseases that eventually afflict nearly every senior. These include diabetes, osteoporosis, and various cardiovascular diseases, including hypertension. Since cortisol naturally elevates in aging populations, many researchers believe there is a direct link. I'm one of them.

Cortisol can damage specific brain regions, too. One primary target for its wrath is our memory-mediating hippocampus. That's unfortunate because of the region's critical role in our survival. It was essential for our species to forge a relationship between stress and memory in the Serengeti: the ability to recall a stressor is also the ability to remember to avoid it. As long as the stress isn't prolonged, the hippocampus learns very valuable lessons about survival and passes them on to you. Remember the upside slope of the U-shaped curve?

Under conditions of prolonged stress, whether from chronic situations or from living past thirty, everything changes, and the hippocampus begins living with a brewing sense of its own demise. Recall that the response is finely tuned only for stresses of short duration. When too much cortisol hangs around too long, it can actually whittle away at hippocampal tissue, causing the organ to atrophy. Some neurons die, meaning excess stress literally causes brain damage. Those neurons that don't expire can lose their ability to connect to one another. Some fail to respond to external signals, and the most alarming failure is the one I've mentioned: the hippocampus increasingly loses the ability to turn off your lifesaving cortisol elevations after the threat has gone away. The thermostat is becoming unresponsive as a direct result of cortisol overexposure. The net result? *More* cortisol overexposure, meaning more damage, which means more cortisol . . . you get the picture. As you age, your brain can turn into the dysfunctional furnace of *A Christmas Story*. This is the downside slope of the U-shaped curve.

How might this show up? You might find yourself more irritable. You might start to lose interest in things, or have unusual bouts of memory loss. Or you might not feel a thing. I wish I could give you clear signs to tell whether you're under the kind of stress that causes brain damage, but I can't. You may have certain genes for resiliency that researchers are beginning to identify. Your brain may become aware of the losses and begin compensating. Behavioral predictions are very hard to make.

Another primary target of cortisol's aggression is the prefrontal cortex (PFC), that vital brain region involved in planning, working memory, and personality development. Prolonged stress destroys the dendrites and spines of specific nerve cells (called pyramidal cells) within discrete layers of the PFC, trashing their connections. It's a massacre. Some experiments show a 40 percent loss of synaptic interactions reaching into the PFC from cortisol overexposure. This results in working memory loss and damage to "higher functions," including personality maintenance. The bad news, it turns out, is quite bad.

It gets worse. The amygdala, which governs your primitive emotions, is supposed to act like a chained beast, shackled to a strong, well-functioning PFC. With the PFC increasingly out of the picture, your brain shifts to a sustained, emotional state of "fight or flight." Your emotions appear to be losing their governors. This is because the amygdala and associated regions don't suffer nearly the amount of damage the PFC or hippocampus do. In fact, the amygdala appears to get bigger and its internal architecture more complex with increasingly chronic stress. So, both socializing and stress—nurture and nature—can increase the size of the amygdala. It's unclear, in the case of stress, whether a bigger amygdala is good or bad, or how it changes behavior.

It is time for us to revisit the good news side of this story.

As I mentioned at the beginning of this chapter, seniors actually *feel* less stressed than their younger counterparts. How can this be? Here are some speculations.

We know that when seniors are shown disturbing photographs, their amygdalae don't overreact like the amygdalae of younger people. This may explain why seniors pay less attention to negative information than youngsters, and they don't remember the details of the aversive material as well, either. It's possible that seniors just don't get as upset with environmental stimuli, even when awash in hormones, because of their changing amygdalae. This may result in the happiness uptick we explored in the last chapter.

It's also possible that the brain's adaptive ability is kicking in. The brain is aware of the internal changes it experiences from aging, and it sometimes tries to correct for them. We'll see a powerful example of the brain responding to loss when we discuss memory. In the case of stress, it's possible that the brain detects age-related changes in stress hormone biology and summons specific compensatory processes to deal with them. You must remember that Old Man Parker, profanity aside, actually got his *Christmas Story* furnace working. It functioned fine for the rest of the movie.

We also know that how stressed seniors feel about aging can actually change the way their brains age. Consider the concept of "age identity," a subjective opinion about how old you think you are (versus how old you actually are). People who feel younger than their chronological age do better on cognitive tests than those who feel older. The magic number appears to be twelve. If your subjective age identity is twelve years younger than your actual age, the improvements really spice up your cognitive scores. Who knew that there would be so much neuroscience to support the quote of noted author Gabriel

García Márquez, who was still writing at eighty-one: "Age isn't how old you are but how old you feel."

Researchers are uncovering more good news about the aging stress response. Recall the cortisol-mediated erosion of the hippocampus we discussed earlier. The damage is not permanent, for the hippocampus is capable of making new neural tissue from resident pools of progenitor cells. This process is called neurogenesis, literally "creating neurons." With new neurons comes improved memory. We'll talk more about how to aid this process in the exercise chapter. Though cortisol can damage the hippocampus, the brain is capable of fighting back. And it can put up its dukes at any age.

Wait, females are a different story

We have one last important issue concerning stress to consider. To describe it, we turn to experiments done by a consortium of researchers in Canada and the United States.

The scientists were studying stress responses in mammals, specifically anxiety and pain in rats and mice. Anyone who does this kind of research routinely knows that, instead of seeing clear patterns in the data, you often obtain a wide variety of stress responses—even when controlling for every variable you can think of. How maddening. The consortium of researchers may have discovered one of the reasons, and it's both welcome and troubling.

One variable labs don't usually take into account is the sex of the human experimenting with the animal. Someone in the consortium decided to control for exactly that, and got a big, fat, disturbing finding. Rats of both sexes detected the sex of the researcher in the room. They changed their stress responses depending on whether the researcher was male or female.

You did not read that wrong. The rats reacted differentially to the sex of the human, and spoiler alert: it doesn't look good for guys. If the experimenter was male, animal stress responses increased (about 40 percent above baseline) during the experiments. If she was female, the animal's stress response *decreased* (yep, below baseline). Turns out the rats were responding to the underarm sweat of the human, which differs in chemical composition between men and women.

These results were met with astonishment, applause, and concern. Sex-based issues are often not taken into account in behavioral work. Yet this experiment clearly shows they should be, down to the sex of the human working with the animals. The research world has had to recalibrate many of its findings concerning stress, as you would expect. You might wonder if differences exist in how older men and women respond to stress as their brains age. While research in the area could use a fresh infusion of funds, the answer is tentatively yes. Three findings are worth highlighting.

The first concerns changes in hippocampal volume, which supposedly shrinks with age. Take sex into account, and a different picture emerges. It's mostly the *male* hippocampus that shrinks with age. Female structures contract somewhat, but the correlation with aging is four times as strong with men as women. We don't know if this translates into behavioral differences—yet another great example of why society should be funding more of this type of research.

The second finding concerns behavioral reactions to environmental stress. We now know that elevated cortisol levels affect older women more negatively than men when it comes to both emotional well-being and cognitive ability. Researchers observe this by "challenging" the brain with stressors under controlled laboratory conditions. The challenge can be psychological, like viewing some nasty news video, or biochemical, like consuming a stress-inducing drug. Older males also react to these challenges, certainly, but the reaction is three times greater in women. The reason may have to do with the hormone estrogen. The stress system that uses cortisol (called the HPA axis, short for hypothalamic-pituitary-adrenal) is much more reactive in postmenopausal women than in premenopausal women.

The third finding concerns the prevalence of age-related dementias. Dementia can indiscriminately raid any aging brain, like a marauding band of Vikings, but it tends to prefer female tissue. Alzheimer's is the classic example. According to the Alzheimer's Association, two-thirds of all people diagnosed with the disease in the United States are women. About 16 percent of women older than seventy-one have the disease, compared with 11 percent of men in the same age group.

Why is dementia so sexist? We used to think it was simply because women lived longer than men, which has an internal logic, since age is the prima facie predictor of any dementia, including Alzheimer's. We don't think that anymore. There appear to be sex-based, even genetic, reasons for this difference, and, again, the culprit may be related to estrogen. In some cases, estrogen appears to provide a powerfully protective firewall against the biochemicals that normally give Alzheimer's its potency. The idea is that when estrogen is depleted, the firewall collapses. We will explore these issues in greater detail in the chapter on diseases of the mind.

We can now turn to a more positive subject: an extraordinarily helpful intervention—which, happily, appears to work equally well for both men and women.

Being mindful of mindfulness

Bespectacled Jon Kabat-Zinn seems an unlikely person to spark an international movement. Looking more like an accountant than a global rabble-rouser, he is soft of voice and slight of build. He speaks calmly, deliberately, with a slight New York accent. But rabble-rouser he is. Active in the anti-war movement in college, he became a leading opponent of Massachusetts Institute of Technology's acceptance of military research funds. He received his PhD in molecular biology at MIT under world-renowned microbiologist Salvador Luria.

While at MIT, Kabat-Zinn began studying Buddhism and yoga. Perhaps in reaction to his scientific studies, Kabat-Zinn came to believe that modern medicine—from research to clinic—was missing something important about the human experience. Combining his meditative practice with his scientific expertise, Kabat-Zinn—now professor of medicine emeritus at the University of Massachusetts—developed a series of techniques referred to as mindfulness-based stress reduction. It is not an exaggeration to say that Kabat-Zinn's ideas revolutionized the field of mind-body medicine, putting it on firm scientific footing.

Now, his technique is one of the most powerful anti-stress therapies ever shown to actually work in the elderly population. That's why I'm making it the crown jewel of my stress reduction recommendations. I advocate a healthy, daily exposure to mindfulness, as long as you're careful about what type of mindfulness you practice.

If that last sentence sounds like a warning, you're reading it correctly. Mindfulness has become a darling of pop culture in recent years—it even made the cover of *Time*—and is in real danger of being watered down, inaccurately described, or both. (A quick "mindfulness" search on Amazon can pull up more than a thousand titles on the subject, including mindfulness for your dog!)

But as long as we choose to ingest only peer-reviewed findings, we should be in good shape. I will define some basic terms, quote directly from Kabat-Zinn, then implore you to explore the reference section on our website. There you will find how-to versions of protocols that have been tested in rigorous, well-characterized trials. If you'd like to practice mindfulness-based stress reduction—and I encourage you strongly to consider it—reading about evidence-based practice would be a great place to start.

Let me give you the CliffsNotes.

Mindfulness, put simply, is a series of contemplative exercises that gently and nonjudgmentally ask you to focus your brain on the *now* rather than on the past or future. Kabat-Zinn puts it this way: "Mindfulness means paying attention in a particular way; on purpose, in the present moment, and nonjudgmentally."

The training exercises have two large components. The first is awareness of the present. Mindfulness invites you to pay attention to the intimate details of whatever is happening at this moment, excluding all else. You start in the physical realm. Concentrating on your breathing is a popular first exercise. So is focusing on a body part, like the sensations in your left foot. Letting raisins linger in your mouth is popular, too. Whereas some meditative styles ask you to clear your mind, mindfulness does the exact opposite, asking you to fill it. With *focus*.

The second component is acceptance. Mindfulness entreats you to observe your present-moment experiences without judging them. It's a method of asking you to observe your life without getting into a quarrel with it. This means not demanding that certain thoughts, emotions, or sensations change—or even go away. At the present moment, they just are. Awareness and acceptance of the present are the two key components common to every definition of mindfulness used in research. These are the ones we'll use, too.

Mindfulness meditation is simple but not easy. Consider the distractions typically experienced by a rookie student. The instructor asks the class to do some breathing exercises, then focus on their foreheads, in this fictional account:

> Okay, focusing on my forehead—focusing on my forehead. Ah, hello forehead. Wait. I didn't take out the trash. Why won't my husband take out the trash? Do I look like a—oh no, focus on your forehead, your forehead. Breathe in. I'm focusing on my forehead. Yikes, my stomach is growling. Can anybody hear? Embarrassing! And hungry. Loved that salmon yesterday, but I poured that stupid butter sauce all over it. Why do I always do that? Okay, can't judge that. Back to my forehead. Breathe out. Gentle. Glad the headache in my forehead is gone. I wish my boss was gone. Is that why I get headaches? She's such a petty—whoops, where's my forehead? Don't beat yourself up. Just back to . . .

It reminds me of a poster where a serene-looking woman practicing meditation says: "Come on, inner peace, I don't have all damn day!"

No question our busy lives don't take naturally to mindfulness. But if we stick with it, really good things will happen to our brains. These good things fall into two categories: emotional regulation (especially the ability to manage stress) and cognition (especially the ability to pay attention).

Put bluntly, mindfulness calms you down. This has all sorts of behavioral consequences. Seniors who practice mindfulness sleep better than those who do not, for example, probably because of lowered cortisol levels. Mindful seniors show marked reductions in depression and anxiety. They report ruminating less frequently over negative things. People who practice mindfulness don't feel as lonely, either, and describe sometimes dramatic changes in the amount and quality of happiness they experience daily.

Though this has not been measured directly, some researchers believe mindfulness extends life. This is not a low-watt assertion. They point to studies on its effects on both immune and cardiovascular systems. Seniors who practice mindfulness suffer fewer infectious diseases. And mindful seniors are 86 percent more likely to score in the very positive range for markers of cardiovascular health than those who aren't practicing. Given that immune dysfunction, heart disease, and high blood pressure are associated with early death (depression is, too), these scientists may be on to something.

Mindfulness also has positive effects on cognition. Attentional abilities are elevated the most. Quoting from a review article: "The strongest finding was significantly enhanced attention (e.g., lower stimulus overselectivity, increased sustained attention and significantly smaller attentional blink) after mindfulness-based meditation practices. There was also evidence that meditation may improve overall cognition and executive functions."

These data fairly laugh with optimism, and I'd like to explain one of the findings in greater detail. Attentional blink describes the awareness lag you experience when your brain changes tasks. It takes time to switch tasks, five hundred milliseconds or so—about the length of time it takes to blink your eyes. As you get older, it takes you more time to switch between tasks, and the blink is longer. Unless you give your older brain mindfulness training. Then you get improvements of about 30 percent over the brains of people your same age who didn't get mindfulness training. That's almost the same amount of improvement you get over the brains of untrained twentysomethings!

This is a big deal. Mindfulness alters the ability of the aging mind to allocate its attentional resources, making the mind more efficient. As we'll see later, the aging brain experiences a marked decrease in its ability to sort through sensory information effectively. Mindfulness is a big help.

Attention isn't the only cognitive talent affected. Positive changes have been noted in visuospatial processing, working memory, cognitive flexibility, and verbal fluency. See why my endorsement is so strong? The twin ideas of awareness and acceptance can literally rewire your behavior. And, as we'll see in a moment, rewire your brain as well. To understand these mechanisms, we need look no further than NBA legend Phil Jackson, a man well acquainted with success in his later years.

Full-court press

Jackson, the former NBA coach who led the Chicago Bulls to six world championships and the Los Angeles Lakers to three, may be America's most famous devotee of mindfulness. He's advocated it in his book *Sacred Hoops*, using words seemingly straight out of a Kabat-Zinn playbook: "In basketball, as in life, true joy comes from being fully present in each and every moment, not just when things are going your way." Some quotes are more cryptic (though still meditative): "Not only is there more to life than basketball, there's a lot more to basketball than basketball." Then he has sayings as down-to-earth as an offensive rebound. "If you meet the Buddha in the lane, feed him the ball!" Jackson has often been asked to come out of retirement, which he's done a few times. In 2014, at age sixty-eight, he received a $60 million contract to become president of the New York Knicks. While that position didn't pan out, ending in 2017, Jackson is still ranked as one of the greatest coaches in NBA history. Jackson credits such success to his emphasis on the most critical planks in any athlete's court: the mental ones.

Researchers would agree. Many labs study the neurological mechanisms behind mindfulness training, and not just in athletes. What exactly is mindfulness doing to reduce stress and increase attention? You might guess cortisol would be a favorite investigative target. Good guess. Lower cortisol levels are clearly part of the stress-reduction story. But not the whole story. Researchers had mixed results replicating critical components of these cortisol data. So they looked elsewhere, hypothesizing that mindfulness-based stress reduction changes the function of the amygdala. Another good guess.

You remember the amygdala, that mighty emotion-generating powerhouse smaller than a hangnail? When people who practice mindfulness are shown distressing environmental stimuli (like slasher movies), their amygdalae show reduced activation compared with untrained controls. The resting state of mindfulness pros is at a lower base level, too, which suggests regular practice of mindfulness leads to a general overall calming. Though the behavioral effects are clear, we're just beginning to understand the molecular machinery behind its effectiveness. Exactly how cortisol regulation and amgydalar changes relate to the observed stress reduction is under active investigation.

Attention on attention

Emotion hasn't gotten all the attention. Attention has, too: What exactly is mindfulness doing to improve focus? One fruitful effort involved studying a neurological area that sounds more like a sports conference than a brain region—the ACC, short for anterior cingulate cortex. The ACC is a medium-size neural subdivision located several inches behind your forehead, just above the eyes. It has many functions, from maintaining attentional states to maintaining a mental gadget we call executive control. It's also involved in error detection and problem solving. The ACC does these latter activities using one of the brain's best-named neural bundles, the von Economo neurons (saddled now with the much more boring title "spindle cells"). Such specialized cells are found only in the world's smartest animals, like elephants, apes, certain whales, and, of course, you and me.

Mindfulness affects attentional states by continually *activating* these smarty-pants regions, including Economo's spindly cells. The regions are turned up to a higher level in mindful people compared with untrained controls—and stay that way, even when the mindful folks are in a resting state. This activation may affect the structure of the brain. There's more white matter wrapping the neurons of these regions in mindful seniors. Remember white matter, that wonderful neural insulation we discussed in the intro? It helps create efficient electrical signaling in neurons lucky enough to have these myelin sheaths. It is quite possible that mindfulness exerts its effects on the brain by strengthening, and thus rewiring, certain "on switch" parts of the ACC.

How might the ACC work in concert with the amygdala and cortisol levels? Several labs have dedicated themselves to charting out the circuit diagrams behind mindfulness. Most of their efforts outline a single fact: it will be a long time before we have the charts. This is exciting on many levels, from the happy fact that there are still many frontiers to conquer to the equally happy fact that people like me will still have a job decades from now, years after retirement age.

Just like Phil, minus the $60 million.

Of mice and men

Here's a sad story. It's a bit of an object lesson concerning mindfulness, plus the advice in the first chapter to have lots of friends and probably every other recommendation in this book.

Flowers for Algernon is a science fiction novel. I read it as a kid and will never forget it. The story concerns a mouse named Algernon and a janitor named Charlie. The mouse has a typical level of rodent intelligence; Charlie has an IQ of 68. Both are selected to undergo a surgery in an attempt to make them smarter. The surgery works. Algernon sails through the standard laboratory intellectual benchmarks. Charlie's IQ soars to over 180.

After a while, however, it becomes clear the boost is only temporary. The mouse deteriorates first, and eventually dies. He is placed in a small coffin and is to be buried in Charlie's backyard. The former janitor's brain begins deteriorating soon after, regressing to presurgical levels. It's a cruel effacement: Charlie is left with the memory that he used to be smart. Poignantly, Charlie's last request is for someone to buy flowers to put on Algernon's grave. Cue the Kleenex.

Why do I bring up this depressing story? In this book, I am discussing lifestyle changes that, if followed, will statistically allow you a more comfortable ride on the aging train. But note how I couched it: *lifestyle* changes. Not temporary fixes that you can use like a Band-Aid until the "owie" goes away. This owie—the aging process—is never going away. That means the changes you make in your lifestyle should never go away, either.

Evidence for this unpleasant admonition comes from a study in which students visited residents of a nursing home once a week. The residents were divided into four groups. In the first group, the student dictated the time of the visit. In the second group, the resident chose the time. The third group received random visits: they didn't know when the student was coming; it just averaged once a week. The fourth group received no student visitors. Various mental and physical assessments were given to the residents throughout.

As you might predict from the chapter on friends, the residents who had the social interaction did much better than those who did not (in mood, health, cognition, etc.).

But then, like *Flowers for Algernon*, the story turns very sad. After the visits ended, the researchers continued to measure how the seniors were doing. The seniors who had received regular visits started faring worse over time—much worse—than those who had never received visits. And much worse than their own baselines, taken before the experiment had begun. The visits, if kept up, made them smarter, healthier, and happier. But once the visits stopped, their brain functions regressed *below* pre-experimental levels.

One way to interpret these findings is "Maybe it would have been better if they never had any extra social interactions." Another is to say, "Make sure that social interactions are a permanent part of a senior's day." This is what I mean by lifestyle changes. There are great reasons to fear what happens if you don't create a robust social schedule for the rest of your life or practice mindfulness meditation for the rest of your life. There are equally great reasons to be exhilarated if you do.

SUMMARY

Mindfulness not only soothes but improves

- Stress is biologically intended to keep you out of danger. It is supposed to be a temporary state. Stay stressed too long, and it becomes damaging to your brain's systems.
- Strive to be positive about aging. If you feel young, your cognitive abilities improve.
- Practicing mindfulness consisting of contemplative exercises that ask you to focus your brain on the present, rather than the past or future, can both reduce stress and boost cognition.
- Improving your lifestyle choices needs to be a consistent and active part of your routine if you want to enjoy both physical and cognitive benefits as you age.

your memory

brain rule

Remember, it's never too late to learn—or to teach

God gave us memory
so that we might have roses in December.
—James M. Barrie

Not only is my short-term memory horrible,
but so is my short-term memory.
—Anonymous

THE TITLE OF THE following true story should be "Amazing Wife to the Rescue."

I was once introduced to a most engaging fellow at a Seattle reception, and we were soon lost in the thick fog of scientific conversation. My wife, finishing a chat with a friend, started walking my way. I knew proper introductions would soon be in order, and I was immediately confronted with this paralyzing, embarrassing fact: I had completely forgotten my new colleague's name! My wife glanced at me upon arriving, sensed that my social memory was stuck in tree sap, held out her hand first, and voluntarily introduced herself. The gentleman quickly returned the favor. See? Amazing Wife to the Rescue.

Forgetful moments like these are painfully common as we age, and they only become more frequent. Comedian George Burns riffed on them famously. "First you forget names. Then you forget faces. Next you forget to pull your zipper up, and finally, you forget to pull it down!" he quipped, in one of his ageless routines about growing old.

Burns's lively flippancies are a terrific example of memory systems that stay vibrant well into old age. How does that jibe with Amazing Wife to the Rescue–style forgetfulness? Our brains have multiple memory systems, and, as we'll discover, they don't age at the same rate. So which changes should keep us up at night? Which can we safely ignore? Is there anything we can do about the memory systems we start to lose?

These are the questions we'll tackle in this chapter, starting with what typically happens to memory as we age. Spoiler alert: we're going to be doing a fair amount of myth busting.

Many types of memory

As you know, it's a mistake to think there's a solitary memory system in the brain, as if a single hard drive were embedded in our foreheads. Rather, there are multiple memory systems within the brain, almost as if the organ were a fancy laptop with twenty or thirty separate hard drives.

Each system is in charge of processing a specific type of memory, each composed of freelancing neural circuits working in a semi-independent fashion. As an example, suppose you remember a high school shop class where your friend Jack got cut while you were all learning how to use a lathe. Learning how to operate a lathe prior to the accident involved a specific memory domain (motor). Recalling that the person who got cut is named Jack and not Brian uses another domain of memory (declarative). Recalling that you watched it in time and space—morning shop class—complete with a cast of characters, you and Jack, uses still another memory domain (episodic).

These systems talk to one another constantly, integrating and updating their findings in tiny fractions of seconds. Yet how they do this is mostly unknown. We mentioned that it's more complicated than a reel-to-reel tape recorder with a playback button. And just to make matters more complex, we have both short-term and long-term forms of memory systems. For simplicity's sake, we'll be focusing on the long-term varieties, except where indicated.

Given how much scientists don't know about memory, any overarching framework attempting to organize it will have major theoretical fracture lines. But a framework I like—the one we'll use here—involves organizing human memory by whether conscious or unconscious functions are stimulated when processing specific types of information.

One system that involves conscious retrieval is called declarative memory—named for memories that are easy to declare. Declarative memory has two components: semantic memory (which allows you to remember the Pledge of Allegiance) and episodic memory (which allows you to remember what happened in *Gilligan's Island*). What do I mean by conscious retrieval? Say I ask you how old you are, and you respond, "None of your business." You know your age, which you bring up consciously. You use your knowledge of English to respond to my question in an indignant manner, also consciously.

And then there are learned skills that you call up without really being aware of it. Take driving a car. Do you consciously retrieve the skills from long-term memory and whisper to yourself: "I will now open the driver's door, get into the left seat, grab the key between my thumb and index finger, insert it into the ignition, turn it thirty degrees clockwise, and wait for the engine to start"? Of course not. You simply get in the car and drive it, awareness mostly not required. This type of memory is called procedural memory. One of the distinguishing differences between procedural memory and declarative memory is conscious awareness.

Let's be clear: all memory systems, conscious-tinged or not, are formed from learned experiences. You weren't born offended by rude questions any more than you were born able to drive. Nonetheless, these phenomena engage different parts of the brain as you learn them. To describe this variation, we scientists puff out our chests and stuffily declare, "Memory is not a unitary phenomenon."

And neither is the aging of those memory systems. George Burns, the comedian we met earlier, is a good guy to help me explain. He signed a lifetime contract to do stand-up at a Las Vegas casino.

At the tender age of ninety-six.

Oh, God! It's George Burns

"You know you're getting old when you stop to tie your shoes and wonder what else you can do while you're down there," the comedian quipped. He joked about his fifteen-cigar-a-day habit: "At my age, I have to hold on to something" and about sex being like trying to play pool with a rope. "I'd go out with women my age," he said, "but there are no women my age." He was asked to portray the Almighty in the popular *Oh, God!* series of movies. When asked how the casting director selected him for the title role, Burns joked, "I was the closest to Him in age." He'd already won an Academy Award at age eighty. Such was their belief in his vitality, the executives at Caesars Palace inked a contract with this funny ninety-six-year-old just so they could have broadcast rights to his hundredth-birthday performance. Here's why his comedic instinct was still going strong.

Semantic memory, a memory for facts, doesn't erode with age. Access to its supporting memorized database—your vocabulary—actually *increases* with the passing years. You score a 25 on performance tests in your twenties. By the time you reach your late sixties, your score is just north of 27 (!). That might not sound like a big deal, but given the elderly brain's reputation for memory loss, few people expected it to show increases. Yet that's exactly what is observed.

Procedural memory isn't priced for quick sale, either. Procedural memory (nonconscious retrieval, remember, falling under the umbrella of motor memory) remains steady as the years go by, although some studies also demonstrate a slight improvement. One experiment, for example, taught a visuomotor task to both younger and older populations, then tested memory performance two years later. Motor memory, as measured by mean performance times, improved

10 percent in younger populations. When the older populations were measured, their motor memories had improved 13 percent.

The fact that these types of memories stay robust over the years adds up to good news: you really do get wiser with age, depending on how you define wiser (and age). These findings come from the obvious insight that we seniors have brains chock-full of experience, which provides two measurable benefits: First, older people have access to a larger fund of knowledge. This gives us a broader array of options for decision making. That's handy when the issues are as complex, confusing, and nuanced as a Middle East peace process. Or our adult children.

Second, our decision making is less impulsive, more thoughtful. It takes longer, simply because we have more options to weigh (it's the load of all those extra memory traces). Senior brains are still flexible and plastic, but decision making becomes more metabolically costly to the brain the more stuffed it is with information. The upshot: seniors tend not to make stupid mistakes. Here's how one paper describes the phenomenon: "The brains of healthy older adults are less likely, and may have less need, to react to environmental challenges with a plastic response than the brains of children and adolescents. In other words, older adults have a richer model of the world that enables deployment of established behavioral repertoires."

You might call this richer model, as some researchers do, "wisdom."

Here again, George Burns's life is instructive. From vaudeville to radio, television to movies, Burns was one of the few comedians to work in every entertainment medium the twentieth century offered. By age ninety-six, his brain had grown fat from the accumulated wisdom of almost eight decades of steady work.

No wonder they asked him to play God.

Now for the bad news

Not all memory systems are preserved with age. One type that declines might best be illustrated not by an old comic but by an old Pixar movie.

Our family has always loved Pixar's delightful *Finding Nemo*. In the story, Nemo's dad (a clownfish) watches his son get kidnapped by a group of divers. Dad runs smacks into Dory, a cobalt blue tang fish voiced by Ellen DeGeneres. She excitedly reveals she's seen the divers' boat, exclaiming: "It went this way! Follow me!" They both start swimming furiously in the direction of the wake.

But not for long. Dory soon slows down, then meanders in a zigzag, looking back at Nemo's dad in an increasingly suspicious manner. She doesn't seem to recognize him anymore. "Will you quit it!" she exclaims, turning on the suddenly surprised clownfish. "What are you talking about?" Nemo's dad exclaims. "You're showing me which way the boat went!"

Dory stops and suddenly smiles: "Hey, I've seen a boat. It passed by not too long ago." With the vigor of an igniting rocket, her brain reenergizes. "It went this way! This way! Follow me!" She blasts off in the same direction as before. Nemo's dad, frustrated, confronts her head-on about this obvious memory lapse, and they stop swimming. "I'm so sorry—I suffer from short-term memory loss," she explains. "I forget things almost instantly. Runs in the family."

That's a terrific example of a cognitive work space scientists call working memory. We used to call it short-term memory, believing it to be a simple, passive storage locker for temporarily storing information. But that was only a long-distance cousin of the truth. We still think it's a temporary work space, but one that is nowhere near simple. Or passive.

Alan Baddeley is the British researcher who first coined the term "working memory." He posited that this work space was dynamic, composed of subprocesses, functioning like shifting piles of manila file folders atop a busy office desk. He was right on all counts. One folder in the working memory work space is useful for temporarily holding visual information (the visuospatial sketch pad). Another folder is useful for temporarily holding verbal information (the phonological loop). Still another folder is in charge of coordinating all the others, appropriately termed the central executor. This last subprocess doesn't hold anything except a program that keeps track of what the others are doing.

Working memory deficits can show up in the most embarrassing ways. You begin to lose your keys more often. You forget what you were about to say, or do, or you lose track of what somebody else says or does. You mention something to a friend, only to have them stop you and say you'd told them before. We've all had these experiences. The decline can be dramatic. One research paper showed that in our twenties, we score on a normalized scale for working memory at about 0.6 (for the particulars of the test, see the references on brainrules.net). That's pretty high. As we get older, unfortunately, the numbers go south. At age forty the score is about 0.2 (not so high) and by age eighty, it has plummeted to a −0.6 (really low). Forgetfulness settles on our brain like a net floating down from above. Working memory is part of a larger network called executive function (EF), which experiences a decline I'll spend some time detailing in a later chapter. Suffice it to say, working memory dysfunction, Dory's enduring—and endearing—quality, will eventually gift us the same way, too.

By the way, Dory was right. Working-memory abilities really *do* run in families. Which means if you want to preserve it, you must choose your parents wisely. Or, short of that, follow the suggestions in this book.

I'll have much more to say about what to do, but right now, I need to deliver some more bad news. It involves one of the most famous prizefighters who ever lived.

Down for the count

Short-term memories aren't the only ones swimming in troubled waters. Certain long-term-memory gadgets encounter turbulence, too.

A case in point is nicely illustrated by an episode of an old TV show called *This Is Your Life*. It starred one of the most celebrated athletes of all time, Muhammad Ali, the late, great, TV-friendly boxer as famous for his mouth as for his fists. And for his self-confidence, bright with quotes like "I'm so mean, I make medicine sick" and "I should be a postage stamp. That's the only way I'll ever get licked."

This Is Your Life functioned as part biographical sketch, part interpersonal ambush. It often featured famous guests, and the hook was to surprise them with appearances from people in their past, flown in just for the show, some not seen for decades. Ali's 1978 spot featured visits from his parents, brother, wife, and other legendary boxers. And one particularly moving segment, a taped interview from legendary entertainer Tom Jones (Who knew they were friends?), in which he recalls when they first met.

"I'm sitting here in Las Vegas, in the dressing room, between shows . . . the time, I think it was about ten years ago, in the Latin Casino, in Cherry Hill, New Jersey, knocking at the door, I looked up, and you were standing there . . ." The most amazing part is Ali's reaction as Jones begins speaking. He looks stunned. As Jones continues, Ali wipes his eyes and nose. "And we've been friends ever since," Mr. Jones finishes. Ali just sits there for a moment. In a life ringed with past glory, the champ appears KO'd not by an opponent but by a memory.

Episodic memory, that lively, previously mentioned subdivision of declarative memory, is just what it sounds like. It is a memory for episodes, the information about events occurring in a certain context and—this is important—interacting through time. Casts of characters are usually interacting in these events. If the character happens to be you, we call it autobiographical episodic memory. Episodic memory is in charge of answering questions like "what, where, and when"—standard fare in a typical *This Is Your Life* installment.

Episodic memory marries two components: the information being retrieved and the context in which the information is recalled. The former is probably just good old semantic memory—the memory for facts. But the latter is unique to episodic memory and gets its own name: "source memory." Think of it like a person giving a speech. Semantic memory recalls the content of the speech. Source memory remembers who said it.

Though episodic memory dips into the deep wells of your semantic reservoirs (Ali is not an "episode," after all), episodic memory in the brain is structurally distinct. How do we know this? Some people are born with ridiculously powerful episodic memories, but they have average-to-poor semantic memories. One famous case concerned a woman who remembered virtually everything that had happened to her since childhood, without fail and without error. Her autobiographical episodic memory appeared flawless. Yet she was a below-average student in school. She had a fairly hard time memorizing run-of-the-mill facts, and she had to make lists to remember certain mundane things. Her declarative memory was flawed. She could remember exactly what she had for dinner eight years, seven days, and four hours ago, yet could not remember her times tables. Distinct systems, indeed.

Now for the bad news: episodic memory, like working memory, gets worse with age. Research shows there's a 33 percent drop in the ability from your twenties to your seventies. (The peak is around age twenty.) A grandfather has a much harder time remembering what he had for breakfast than his granddaughter does.

We even know the type of information that goes down for the count: source memory. One test measured younger and older test subjects as they watched people giving speeches. The subjects were later asked to recall the content of these speeches, then to match which content was said by which speaker. Older people and younger people could identify the content just fine (semantic memory), but older people had much greater difficulty in identifying who said it (source memory). They failed even to remember the gender of the speaker, a much less cognitively taxing task called partial-source memory.

What's going on with episodic memory from a neurological perspective? Episodic memory involves electrical connections between the hippocampus, which we've discussed before, and something called the default mode network (DMN), which we haven't discussed before. Hippocampal involvement makes sense; it helps mediate many types of memory. DMN involvement will, too, once you know a bit about its functions.

The DMN is a group of far-flung neural networks: regions behind your forehead connecting to regions arcing between your ears. It's called "default" because it's active when you're inactive—bored, daydreaming. The DMN is deeply involved in episodic memory, too, particularly with neurons on the right side of the prefrontal cortex. It makes sense that the neurons generating daydreams might also help construct narratives, as both involve episodic features.

As we age, both the hippocampus and DMN begin to erode. You can see this structurally (volumetric loss) and functionally (connectivity changes). It's here that things get nightmare-like. The brain cannot marshal enough forces to overcome it. Unless you do something intentionally, these changes will become permanent. A moderate loss is typical—happens to everybody—but a severe loss is not. It's one of the signature features of Alzheimer's disease.

Unfortunately, working memory and episodic memory are not the only systems suffering age-related decline. I bet you've experienced the third one already.

It's on the tip of my tongue

Two older married couples are walking home from a movie, goes an old joke. The women are chatting in front, men tarrying in back. One man announces, "We went to a really fine restaurant last night. You should go." His friend replies, "What was its name?" The man begins to respond, then becomes frustrated. "I'm afraid I don't remember," he says. "What is the name of that lovely flower everybody likes? You know, the kind you give on Valentine's Day?" "Do you mean a rose?" his friend says, puzzled. "Yep, that's it," says the man, and then calls out to his wife walking ahead.

"Rose? Hey, Rose! What was the name of that restaurant we went to last night?"

Most everybody I know suffers from a variation of the memory loss described in this joke. You want to recall a word, and you have the distinct feeling it's rolling around in your memory like some invisible marble. But it soon it circles down the inexorable cognitive drain, lost forever until noon the next day, when you suddenly recall it. This is the Tip of the Tongue phenomenon (its actual scientific name!). As we age, the frustrating experience becomes more common. On average, Tip of the Tongue irritations increase fourfold when you compare seventy-year-olds to thirty-year-olds.

One of the most intriguing aspects of this loss is what's not lost. In the joke, the older man knew he'd gone to a restaurant, really enjoyed it, and wanted to share that with his friend. And he *did* share something about it verbally, showing his language comprehension was fine. What got him in trouble was finding a specific word.

Here's the upshot. Language comprehension and general word production are well preserved into old age, like canned peaches. Losing access to phonological representation, like fruit that sat in the sun too long, is not.

It's obvious that memory is uneven in its decline. Is there a generalizable time line scientists use to track its aggravating progress? This is an important question. Many seniors fear that the dark shadow of dementia has claimed another few square inches of their brains every time they can't remember their favorite wine. Fortunately, most of these memory losses are normal and not indicative of anything except a large backlog of birthdays. And there are things you can do to slow—even reverse—the decline. Only in a few cases are such losses indicative of something more serious, like dementia. We'll describe in a later chapter how to tell the difference between the typical and the terrifying.

In the meantime, you might find perverse comfort in knowing there's lots of disagreement in the scientific community about exactly what declines, by how much, and when. The problem? Aging is so very individually experienced. Pair that with the fact that scientific understanding of how memory works is incredibly incomplete. To stay on the sunny side of peer-reviewed literature, we're reduced to the following two statements:

1. In any one particular year as we age, a few memory gadgets get worse, a few get better, and a few don't change at all.
2. Most everything declines after age thirty.

Working memory, for example, peaks at age twenty-five for most people, stays steady until thirty-five, then begins its long slow journey into the night.

Episodic memory peaks five years earlier than working memory, then takes the same slow slog as its working cousin.

Contrast these findings with data showing overall vocabulary scores don't peak until you reach your sixty-eighth birthday. That may sound positive, but on closer inspection may also sound contradictory. How can that possibly be—especially when your Tip of the Tongue exasperations become annoyingly noticeable soon after you turn twenty-five? You appear to have a Cadillac database for vocabulary, but your ability to access it seems to corrode to a Model T.

Will these puzzles be solved if we pop open the hood of an aging brain and peer inside its whirring retrieval gadgetry? They might—and so we will boldly go where neuroscientists have gone before. To get some help, we're going to enlist an officer of the USS *Enterprise,* its legendary skipper, Captain James T. Kirk, and a battle he once fought with something named a Gorn. No kidding.

By hook or by crook

The Gorn was a reptilian alien in a really cheesy costume, starring in a *Star Trek* episode titled "Arena." The installment begins with Captain Kirk and the Gorn, previously locked in a space fight over territorial rights, who are suddenly whisked to an alien planet by some advanced race. This race strips Gorn and Kirk of their fancy-schmancy space weapons, then forces them to work out their differences by fighting a duel with just their hands and wits.

Of course Captain Kirk is going to win. He finds, lying around the planet, the basic ingredients needed to create a crude ballistic weapon, complete with small-bore cannon (bamboo stalk), diamond-like projectiles, and the components of gunpowder. He fires his makeshift cannon at his reptilian competitor, severely wounds him, and then decides, Shakespearean-like, not to kill him. It was a lesson in creative workarounds, damn the photon torpedoes, with Kirk to the rescue in all his moralizing glory.

(Discovery Channel's show *MythBusters* tried to replicate the technology described in the episode. They found that the bamboo cannon, no matter how well reinforced, always exploded the instant it was lit. Conclusion: Kirk would have been killed no matter how his weapon was designed.)

You can quibble with the scriptwriter's knowledge of physics, but you can't argue with Kirk's compensatory creativity. And that's what our aging brains are providing, as parts of our memory decline.

An example is syntactic processing, the ability to arrange words into cohesive sentences. Scientists investigating older people's brains found that though the verbal skill did not change, the way the brain went about accomplishing it did.

A younger brain normally accomplishes syntactic processing by activating Broca's speech center. The area is named for a nineteenth-century French physician (who was once denounced as a "materialist and a corruptor of youth"), Pierre-Paul Broca. It's a garden patch of neural networks on one side of your brain, just above your left ear (in the inferior frontal cortex and posterior middle temporal gyrus, left lateralized, for you anatomy geeks). Spoken language flows from two regions there, designated BA 45 and BA 44. How do we know that? If you damage those networks, you can't speak in grammatically correct sentences. Your language sounds like gibberish. Your speech comprehension also suffers.

Like an aging celebrity, these networks begin to fade as the brain gets older, the neural pathways connecting separate areas of the brain slowly losing their ability to communicate with one another. This loss of connectivity often predicts loss of function. And that's what was puzzling to researchers, because syntactic processing is well preserved in the aging brain.

Here's where your brain transforms into William Shatner. It grabs that bamboo shoot and improvises. It senses loss, looks around for brain regions not normally used in language, and starts parasitizing their functions. Scientists have observed two such compensatory changes: First, aging brains begin stimulating neurons on the brain's wrong side (the right hemisphere) during language production, recruiting regions not normally associated with syntactic processing. Second, this recruiting drive extends into the prefrontal cortex, activating certain neurons also not normally associated with language. (This recruitment occurs only when the participant is also performing some task. We have no idea why.)

In addition to co-opting, the brain also reorganizes the electrical relationships between the neurons remaining in the language production centers of our youth. Thus your brain appears to be starring in its own version of "Arena," using material lying around its dusty neural corners to battle back the advances of aging.

Captain Kirk would be proud.

The power of the new

"What's this stuff?" a little boy demands, addressing his brother at breakfast. He's pointing to a bowl of Life cereal. The brother shrugs, "Some cereal. Supposed to be good for you." Neither wants to try it and they push the bowl back and forth. Suddenly one gets an idea: "Let's get Mikey!" "Yeah," the other continues. "He won't eat it. He hates *everything*." They shove the bowl over to their younger brother, Mikey, and watch eagerly. To their complete astonishment, Mikey digs in, relishes the experience, and enthusiastically eats more. "He likes it! Hey, Mikey!" the brother cries, amazed. The screen cuts to a shot of the product and sales pitch.

This thirty-second spot is regularly voted one of the top ten commercials of all time and was responsible for gazillions of sales for the Quaker Oats Company. Though it's hard to believe you could make an indelible impression simply by trying something new—and taking only thirty seconds to do it—Mikey is living proof that you can.

Draw a circle around this signature idea, that trying something new can produce benefits, because it's nearly everything science knows about how to improve aging memory systems.

That's right. Even though memory naturally declines (and most memory types don't have naturally occurring neural rescuers), we're not left hopeless. We can treat the corrosive effects of time with a one-sentence prescription: "Go back to school."

Yes, I am putting on my stern professorial hat, thrusting my finger into the air, and demanding that your brain take up the habit of lifelong learning. Enroll in a class. Pick up a new language. Read until you can't see anything anymore. An aging brain is fully capable of learning new things. To keep that talent healthy, you have to plunge yourself into the deep end of learning environments every day. No exceptions. You take up Mikey's willing spoon and swipe away the cobwebs of age-related memory decline.

Researchers even know the type of learning that's most nutritious. It's based on the psychological concept of "engagement," which has two types. The first is receptive engagement, where you learn things passively, leisurely, stimulating areas of knowledge with which you are already familiar. This has been shown to improve memory in aging populations.

Yet there's a better way. If you want the Energizer Bunny of memory improvement, go for "productive engagement." Here you experience a novel idea and actively, even aggressively, engage it. The best exercise is to find people with whom you do not agree and regularly argue with them. Productive engagement involves experiencing environments where you find your assumptions challenged, your perspective stretched, your prejudices confronted, your curiosity inspired. Productive engagement is one of the clearest ways to keep your memory batteries from draining.

How do we know this works? Consider research examining the effects of productive engagement on episodic memory. Researchers at the University of Texas at Dallas developed a program called the Synapse Project, which included two types of learning: receptive and productive. Seniors were exposed to one of the two conditions for fifteen hours a week, for three months. The productive-engagement group learned a demanding skill such as digital photography or quilting. The receptive group socialized. After a period of time, both styles were found to improve episodic memory—dramatically, actually—but scores for the productive learners went through the roof. In a 2014 article, lead author Denise Park wrote: "The findings suggest that sustained engagement in cognitively demanding, novel activities enhances memory function in older adulthood . . ."

She's being modest. Episodic memory improved 600 percent above those in the receptive group.

Episodic memory isn't the only function that improves with aggressive learning, nor is the Synapse Project the only concept that works. Teaching other people works beautifully, too. Seniors who taught elementary schoolchildren basic skills, like literacy, library usage, or proper behavior in a classroom, showed dramatic improvements in specific memory domains (and other cognitive functions as well). This is consistent with plenty of research demonstrating that one of the most effective ways to keep your brain sharp over a fund of knowledge is to continually teach it to others.

The results of aggressive learning are so powerful, they even reduce a senior's probability of getting Alzheimer's disease, a notion we'll explore in the chapter on dementia. This simply underscores the clarity of the findings. Even if you hate everything, lift your spoon to try something new. It is one of the best experiences you can give your brain.

Saints be praised

Here's another good thing you can give your aging organ. It's illustrated by a quote that is surprising, mostly because of its source: "Stupidity is also a gift of God, but one mustn't misuse it."

That quote was uttered by none other than Pope John Paul II (now Saint Pope John Paul II). This startled me, because I knew it was a gift his mind never opened.

I swear Pope John Paul's brain was as big as the Vatican library. He spoke at least eight different languages fluently (accounts vary) and may have had a working knowledge of dozens more. He had a love affair with music, cutting an album called *Pope John Paul II Sings at the Festival of Sacrosong*, which sold enough to have actually charted (peaking at No. 126). He even hired his own musical adviser when he moved to the Vatican. He was apparently a voracious reader, too, exuding an enthusiasm for books second only to the other great secular love in his life, the outdoors. He was an accomplished hiker, kayaker, and skier, earning the moniker "Daredevil of the Tatras" (a mountain range in Poland) from his skiing buddies before he became the pope. It must have done some good, for he became the second-longest-serving pope in modern history. He died at eighty-four—full of years, controversy, and adulation.

Whether Saint John Paul knew this or not, most of his lifestyle habits were neural fertilizer, directly in line with what science knows about nurturing one's memory, especially if performance is what you're after.

We know, for example, that bilingual people perform significantly better on cognitive tests than monolingual controls. This includes memory, especially working memory, *regardless of the age at which the language is learned*. There's a minor dose-dependent relationship: people who know three languages outscore people who know two, and both score higher than those who know only one. Fluid intelligence, a measure of creativity and problem solving, is better in bilinguals, too.

Language turns out to be a friend with many long-term benefits. Normal cognitive decline is less steep for bilinguals. Same with their risk for general dementia. The onset of dementia is delayed more than four years compared with monolinguals. These associations are robust enough to warrant a suggestion: when you collect your first Social Security check, use it as tuition to enroll in a foreign language class.

Another saintly example involves exposure to music—even for those with little prior experience outside of listening to the weekly Top 40. One experiment took musically naive seniors and exposed them to a four-month music-training program. They not only learned to play the piano but also were taught music theory and sight-reading. Tests of executive function (which includes working memory) improved dramatically. Participants were happier, as shown by quality-of-life assessments, including measures of depression and acute psychological stress. The control group for this study experienced "other leisurely activities," ranging from computer classes to painting lessons. The results were clear: it was music that did most of the heavy cognitive lifting.

Voracious reading, another papal habit, also turns out to be good for aging brains and, surprisingly, even better for longevity. One twelve-year study showed that if seniors read at least 3.5 hours a day, they were 17 percent less likely to die by a certain age than controls who didn't read. Read more than that and you increase the number to 23 percent. The reading has to be of books, long form. While reading mostly newspaper articles did something positive, the effects were smaller.

A smattering of other habits, sounding like Pope John Paul's daily to-do list, reveals more memory-boosting treasure. Exercise (mountain climbing with the Daredevil, anyone?) is great for both short- and long-term forms of memory. So is meditating. The usual my-parents-already-told-me lifestyle habits apply here, too, like getting enough sleep, eating healthy foods, and hanging around good people. Plus something your parents didn't know to tell you: staying away from the blue light of electronic devices.

This is where our David Attenborough Amazon River analogy is useful. Many tributaries contribute to improving the flow of our aging memory. Taken together, the effects on cognition generally, and memory specifically, are robust enough to be a formula. The more you lift weights in the mental gym, the more you postpone your otherwise natural memory decline. We even know the rate. Every day you exercise your brain above what you do typically delays that deterioration by 0.18 years.

That's an extraordinary thing to say. And it's a scientific result that has the unusual luxury of being backed by heaven. Or at least by the lifestyle of one of the smartest saints to ever wear papal robes.

A private reserve

Why does such a prefrontal—and full-frontal—educational assault work so well? We think it involves something called cognitive reserve. I'd like to introduce to you eighty-two-year-old John Hetlinger, who will help us explain the concept. Hetlinger is a spry, goofy-looking old man, and—from an appearance on the show *America's Got Talent*—a YouTube sensation. When the judges asked him what he did for a living, Hetlinger replied that he'd been an aerospace engineer: former program manager for a Hubble Space Telescope repair project. The judges sat openmouthed for a moment. And this was only the beginning of their collective jaws dropping to the floor.

What completely dislocated their mandibles was when Hetlinger started performing. Drums ticking out the time, Hetlinger whispered the words "Let the bodies hit the floor" with increasing intensity until he was roaring with the irrational energy of a heavy metal front man, screaming "LET THE BODIES HIT THE FLOOOOOOOOR!!!"

Perhaps channeling Black Sabbath from an earlier era, Hetlinger did a take-no-prisoners' version of the hit song "Bodies" from the metal band Drowning Pool. He immediately got a standing ovation. Later, one of the judges asked him, "Is there a mosh pit where you work?" Hetlinger replied, laughing, "No, but there's a lot of beer."

I can't think of experiences separated by more occupational light-years than an energetic heavy metal performance and repairing the Hubble telescope. And from an eighty-two-year-old! Hetlinger seemed to draw from some mysterious unseeable store of energy, enthusiasm, and humor. Brain scientists would agree, though we don't believe it's mysterious—or unseeable. We call Hetlinger's store "cognitive reserve."

Cognitive reserve was birthed from a concept called brain reserve. Brain reserve is a physical measurement involving (a) overall brain size and (b) a census of how many neurons are still available for work. Cognitive reserve measures your ability to use what brain reserve you possess. It was originally hypothesized to explain the observation that some people recover from brain injury failure quickly, and some not at all. The difference turned out to be the amount of cognitive reserve carried prior to injury. If you could increase it, you were more likely to go through life like John Hetlinger than, say, Ozzy Osbourne.

Research shows that pelting your brain with a rainstorm of productive cognitive experiences—in other words, everything we've talked about in this chapter—fills up the cognitive reserve cistern. You can even measure it. For every year of education experienced, cognitive decline is delayed by 0.21 years. (This rate is remarkably similar to the numbers seen delaying memory decline. How, or if, they're related is unknown.) As lead author Mark Antoniou summarizes it: "Cognitive reserve is defined as resilience to neuropathological damage of the brain, and is thought to be the result of experience-based neural changes that are a consequence of a physically and mentally stimulating lifestyle."

Two prominent mechanisms have been proposed to explain these neural changes, each with their own peer-reviewed cheering sections.

The first has the indelible ink of "nature" stamped all over it. Some people are hardwired with a deep cognitive reserve, and are probably born with it. Such people have certain brain regions structurally different from people with poor cognitive reserve. To increase your chance of recovering from some mental injury, it would be wise to have intact neural populations in your frontal, parietal, and temporal cortices.

The second mechanism has the more washable ink of "nurture" stamped all over it. People who've spent a lifetime in mentally and physically demanding environments are much more efficient at using whatever brains they carry into their elder years. They're also more neuroanatomically "nimble," more flexibly able to create alternative neural circuitry when the originals become injured.

Given these inky prequalifications, you might predict the biological bank would deny your request for additional reserve loans once you've reached a certain age. That's where you'd be wrong. It's settled neuroscientific law that you can take up learning at any age. The only closing costs are that you have to start. Consider these reassuring words from Alzheimer's researchers at Columbia: "Even late-stage interventions hold promise to boost cognitive reserve and thus reduce the prevalence of Alzheimer's disease and other age-related problems."

I'm sure Hetlinger would agree that it is never too late to learn anything. The only bodies that really need to hit the floor are the prejudiced ones saying you can't.

SUMMARY

Remember, it's never too late to learn—or to teach

- The brain's memory is like a laptop with thirty separate hard drives, each in charge of a specific type of memory.
- Some memory systems age better than others. Working memory (formerly short-term memory) can decline dramatically, causing forgetfulness. Episodic memory—stories of life events—also tends to decline.
- Procedural memory—for motor skills—remains stable during aging. Vocabulary increases with age.
- Learning a demanding skill is the most scientifically proven way to reduce age-related memory decline.

your mind

brain rule

Train your brain with video games

I've reached an age where my train of thought often leaves the station without me.
—*Anonymous*

Isn't it funny how day by day nothing changes, but when you look back, everything is different?
—*Anonymous*

FOR FANS OF *I Love Lucy*, the fun product "Vitameatavegamin" is taped to the refrigerator door of their brain. The unusual words came from the episode "Lucy Does a TV Commercial" (well worth a peek on YouTube). Protagonist Lucille Ball is shooting a commercial for a fictional health drink called Vitameatavegamin, and we're watching the rehearsals.

"Hello, friends. I'm your Vitameatavegamin girl!" she begins with a smile, plastic as California. "Are you tired, run-down, listless? Do you poop out at parties? Are you unpopular? The answer to all your problems is in this little bottle!" Lucy holds up the product. "Vitameatavegamin contains vitamins, meat, vegetables, and minerals," she continues, swallowing a spoonful of the stuff.

What happens next is the stuff of comedic legend. The bottle contains alcohol or some other mind-altering substance, and after several takes, Lucy shows signs of mental impairment. Her brain's processing speed slows down. What Lucy can pay attention to shrinks, including her ability to stay on script. Her decision making is being compromised and, speech slurred, she barely makes it through the last take: "Do you pop out at parties? Are you un*poop*ular? Well, are you?" Lucy raises her voice, looks whoozily at the camera, pats the bottle. "The answer to all your problems is in this lil ol' bottle . . . vitamins 'n' meat and megetables and vinerals." She hiccups. "So why don't you join the thousands of happy, peppy people and get a great big bottle of Vita-veetie-veenie-meany-miny-moe!" She spills the stuff on the floor, tries unsuccessfully to pour a teaspoon, then takes a giant swig directly from the bottle. In 2009, *TV Guide* ranked "Lucy Does a TV Commercial" as No. 4 of "TV's Top 100 Episodes of All Time."

Lucy's gradual impairment is more than a delightful media-history lesson. Researchers have found we will all experience age-related declines in several cognitive processes, which Lucille Ball's character attempted to keep steady: processing speeds, attentional abilities, and decision making. There's only the calendar to blame, sadly, no alcohol required.

This might sound depressing, but there's actually reason to be hopeful. Researchers have found these same cognitive abilities are quite amenable to external intervention. You can play computer-based games that can slow—or even reverse—declines in processing speed, ability to pay attention, and decision making. Sort of like watching "Lucy Does a TV Commercial" in reverse. Which, come to think of it, might be just as delightful.

We'll drink to solutions later. For now, I want to walk through what happens with each of these three brain processes.

Noisy cocktail parties

The first issue we'll tackle wouldn't seem out of place to modern computer geeks: processing speed. In the world of cognitive neuroscience, processing speed is the rate at which a person performs a task.

What type of neural tempo is being measured depends on what task is being performed. Scientists measure reflexes using motor processing evaluations. They measure perceptual speed and decision making using cognitive processing evaluations. I'm going to focus on perceptual speed limits, which can be divided into three discrete stages. Let's use a real-life example to illustrate them: Imagine you're at one of those annoyingly noisy cocktail parties people feel obligated to attend, and someone's just horse-collared you into hearing about her favorite grandchild's getting into college. Your first stage is intake—the ability to recognize information and haul it into the brain for further processing. (You might say to yourself, "Oh *that* kid. Molly. I know her.") The second stage is reaction, in which you assess the information's meaning, often including a judgment ("Molly actually got into a college?"). The last stage is actionable response, which involves formulating and executing "what to do about it" plans. (You say outwardly, "That's so wonderful," then exit.)

With age, accomplishing all three steps becomes an increasingly Sisyphean task that's often frustrating because it didn't used to be so hard. Processing speed increases dramatically from elementary years to high school, peaks around the time you start college, then starts sunsetting after graduation. Changes are especially noticeable after age forty. On average, you lose about ten milliseconds of speed for every decade you live past age twenty. That may not sound like much, but it's actually a big deal. The processing difference between high-functioning brains and cognitively impaired brains is only about one hundred milliseconds, depending on the study. On certain tests related to symbol replacement, twenty-year-olds are 75 percent faster than seventy-five-year-olds.

Unfortunately, the painful downslope of this inverted U is as easy to feel as arthritis. When people report that their brains are aging, aside from memory jokes, they often unknowingly reference perceptual speed. There is cause for concern. A decrease in processing speed is the greatest predictor of cognitive decline that exists in the research literature—and the greatest statistical detector of who will eventually need help with daily routines. Though geroscience shows that not everybody experiences this rise-peak-fall journey the same way, it is true that everybody goes through it.

What do you experience? You feel as if your brain is getting stuck in sap. It becomes harder to solve problems, and it takes you longer to finish even when successful. It also becomes increasingly difficult to attend to information in the face of competing influences, like at noisy cocktail parties. You can't lip-read as well, something at which we normally excel.

We know many of the reasons why this occurs in the brain. Much of it can be illustrated by looking at the wires in a typical home, which come in many hues.

Why are wires in your house usually covered with colorful coatings? Besides helping you tell them apart, the coating serves as insulation. Wires need insulation in part so they can send power from one location to another. Without it, electricity acts like a river without a bank. It just floods—diffuses—everywhere, and nowhere, except if you touch the wire. Consider high-voltage wires, which aren't coated with insulating materials. If human hands touch them, death ensues. If flammable materials touch them, fire occurs and then death ensues. It's not a problem most of time. The surrounding air provides ample insulation, as long as the wires stay out of reach. That's why they're so high off the ground. A downed transmission line needs to be treated with the same respect you'd give an angry cobra.

Neurons need to be insulated, too, even though you can't ever get a shock from your nerves. White matter is what insulates neurons. Not all parts of a neural cell need insulation—dendrites and cell bodies and telodendria come to mind—and they show up dull gray under close examination. They're collectively termed, unsurprisingly, gray matter. You begin life as a child with a lot of gray matter. Over time you accrete white matter, a process called myelination. The brain isn't fully myelinated until age twenty-five, which means the brain takes last place in the body's race to finish its post-birth developmental schedule.

Without white matter, neurons act like wires without insulation—which in the watery world of the brain means a loss of signal and a slowing down of relevant cognitive processes. Losing neural insulation explains many age-related declines, including processing speeds.

Nature, nurture, and speed

The story behind white matter and cognitive slowing comes from our familiar family of nature and nurture. On the nature side of the household, structural changes occur in the frontal lobes (the areas behind your forehead) that reduce insulating white matter. The cellular mechanisms behind this loss are known, surprisingly, and worth detailing.

White matter is composed of living cells called oligodendrocytes. They wrap around the neural axon (that long, slender part) like wrapping paper around a cardboard tube. When white matter erodes, it's because those oligodendrocytes die off and unravel from the axons. The brain tries to repair the damage by recruiting replacement oligodendrocytes, but it's not a perfect strategy. With age, the originals are replaced with inferior copies, reducing structural integrity. This compromises the quality of the electrical signaling. Processing slows down.

Another molasses-inducing mechanism comes from changes in a brain region we haven't discussed before. It's the cerebellum, which looks like heads of cauliflower stuck to the brain's underside. It's no sedentary cabbage, however. The cerebellum is involved in movement, its most famous function being motor control. Try sewing with your arms flailing around your body every time you try to thread a needle. That's life without a cerebellum.

Motor regulation is not the only function of this multitalented vegetable. The cerebellum also appears to be involved in language, attention, mood, and processing speed—especially in measurements involving motor tasks (like pressing a button). With age, two changes occur that directly alter that speed. First, gray matter volume within the cerebellum shrinks. Second, connections from the cerebellum to far-flung places like the parietal lobe (roughly the region underneath a wide headband) erode. That's a big deal: the parietal lobe helps integrate information from a variety of senses. These negative changes result in slower processing. When combined with the frontal findings, a remarkably crisp picture emerges about why things decelerate.

In addition, vision and hearing decline with age, which can alter the amount and types of data the brain can process. Medical issues such as thyroid problems and cardiovascular issues can turn brains to syrup. So can diabetes. Even respiratory infections can alter speeds, which may help explain the age-relatedness of the problem, given that weaker immune systems are much more common in the elderly.

Nurture plays a part, of course. Not getting enough regular sleep can slow information processing to a crawl. Ditto with stress. And medications like antihistamines and sleep aids, even certain antidepressants. There's our multisource Amazon River analogy again, forcing the brain to take a muddy meander through its ability to solve problems.

That's processing speed. Now we turn to a feature in which processing speed is deeply involved: attentional abilities.

Intellectual hiccups

One early, blurry Seattle morning, I stumbled downstairs to our basement pantry to retrieve some juice. On the way down, I discovered the postapocalyptic remains of my teenage son's party from the night before. Smiling (sort of), I picked up a few pizza crusts, paper plates, and paper cups—then made a mental note to talk to him.

I paused after reaching the pantry. Like Puget Sound fog, a thick feeling suddenly billowed over me. What, for heaven's sake, was I supposed to get down here? I had completely forgotten. Traipsing back upstairs, I discovered for the second time that morning that we had no juice. I felt the attentional amnesia and laughed out loud.

What happened to my memory? Younger brains can create goals and, despite the headwinds of pesky interruptions, still accomplish them. As our brain ages, the ability to ignore those distractions wears down. Pizza-interfering-with-juice is one more signature cognitive behavior of growing old.

How do we know about this intellectual hiccupping? Scientists use an assay called the counter-task test. Our ability to ignore distractions declines from a high of 82 percent when younger (average age twenty-six) to a low of 56 percent when older (average age sixty-seven). That's what happened in the pantry. Rather than ignoring the pizza-littered war zone to get the juice, I became distracted by it. Interestingly, it's not the inability to focus that produces the problem. Older folks concentrate on tasks just as well as younger ones, maybe even better. It's the increasing inability to ignore distractions.

To be fair, room amnesia (yes, scientists actually have a name for it) can happen at any age. The loss involves something called an event boundary. "Doorways are bad. Avoid them at all costs," says University of Notre Dame psychologist Gabriel Radvansky, who's studied the phenomenon for more than twenty years.

My basement sojourn was an example of a single task, interrupted. What about doing two tasks at once? Such simultaneity is often (incorrectly) called multitasking. Scientists have a better term—divided attention—because what we're really doing is switching between tasks.

It gets increasingly hard for us to switch between tasks with age, especially on a moment-to-moment basis. Sadly, this behavior's been on a downhill slog since our sophomore year in college. It's especially tough when a task demands a high degree of attention.

There are many ways to measure divided attention. One method involves concentrating on a laptop while someone stands offscreen demanding you pay attention to something else. In other words, just like every broadcast journalist you've ever seen. Snippy directors whispering into a TV anchor's earpiece while she's trying to bring you the news is the *perfect* experimental example. The more complex the tasks, the harder it is for older brains to keep up.

Scientists have known for years that true multitasking is a myth. It's impossible for any brain to monitor two attention-rich targets simultaneously. The only way your brain can track multiple targets is to use a task-switching strategy. That switching is what researchers measure. Here's the bottom line: older people just don't do this very well. The numbers are similar to the processing speed data just discussed.

There's no better illustration of this than your grandmother driving a car. When changing lanes on a freeway, she might nearly scrape the car next to hers because she was suddenly distracted by an unexpected slowing of the vehicle in front. She might underestimate the distance between cars when parallel parking or become distracted by the raindrops on the windshield in bad weather. These are all toxic distractions.

Processing speed doesn't help, either. As the brain shifts into a lower, slower gear, it begins to choke on the amount of driving problems it can address. Since no cognitive Heimlich maneuver exists to rescue yourself on the freeway, slower processing becomes a dangerous fact of life. It's the leading reason people quit driving when they get older. You may want to continue to operate a motor vehicle, but your brain has other ideas.

We've covered processing speed and attention. Now we'll discuss a process that involves both of them: decision making.

Not-so-fluid intelligence

Wilhelm Wundt may be the most influential scientist you've never heard of. Though he died in 1920, his insights are still ridiculously influential. In this section we are going to talk about one of those ideas, emotion-based decision making and how it ages.

Wundt didn't start out impressively. Kind of a lonely, scrawny kid, he did so poorly in school that one teacher suggested he become a mailman. Things changed when, by some miracle, he was accepted into medical school. There he revealed a lively interest in physiology and an even livelier interest in the life of the mind. Now fully engaged, he embarked on a sixty-five-year research career on human behavior, a journey so incandescent that he's considered to be the founder of modern psychology. His bright light illuminated the careers of many students, some with their own earth-shaking research, and whom you've probably never heard of, either. These include luminaries like G. Stanley Hall, founder of child psychology, and Edward Titchener, creator of the word "empathy." No kidding.

One of Wundt's gold-medal ideas involves the concept of arousal and the role it plays in emotion-based decision making. If we've a choice between two alternatives, we first evaluate them on the basis of perceived benefit. If our brain is positively aroused by some opportunity, we'll move toward it. If our brain is negatively aroused, we'll move away from it. These simple approach-avoidance choices are critical elemental building blocks upon which we make more complex decisions. It's not the only way we decide things, but it explains a lot. I mention it here because approach-avoidance is so clearly affected by the passing years. Our ability to make emotional decisions shifts like tectonic plates as we age.

This shifting should be familiar ground, because we covered some of this in the middle of chapter 3, with the fake lover from London and the way our motivations change over time from promotion to prevention. Researchers discovered such erosions in emotional decision making are only a subset of a larger loss, however. What really fractures is something called fluid intelligence.

Fluid intelligence, roughly defined, is your ability to persuade your problem-solving talents to come out and play. Specifically, it's the facility to apprehend, process, and solve unique problems independent of your personal experience with them. As one research paper noted, fluid intelligence involves our "abilities to flexibly generate, transform, and manipulate new information."

Since information needs to be held in a volatile memory buffer, at least while you're manipulating it, you might predict that working memory plays some role in the ability. Lab findings would show you're right. Fluid intelligence is highly correlated with working memory ability. They may, in fact, influence each other. And we've already seen that working memory takes a dive with age.

Fluid intelligence is often contrasted with its talented twin, crystallized intelligence. Crystallized intelligence is defined as the ability to draw from material learned by experience, using information previously stored in a structured database. As you'll recall, not all memory systems erode with age (some improve), and you see this statistically with crystallized intelligence. Depending on how you measure it, crystallized intelligence stays fairly stable throughout life.

That's distinctly *not* the case for fluid intelligence. Typical fluid intelligence scores drop almost 40 percent between the ages of twenty (when they peak) and seventy-five. So decision-making abilities that utilize gadgets from the fluid intelligence toolbox erode over time. That includes decisions requiring inputs from a variety of sources *simultaneously*—like placing the many dishes of a rich Thanksgiving dinner on the table without any getting cold. (It doesn't help that working memory, a memory gadget that also erodes, is involved.) Fluid intelligence also includes decisions involving approach-avoidance issues, which means you can insert Wundt's arousal ideas right into this paragraph.

All of this is going on in a neural network that Yale researchers call the affect-integration-motivation (AIM) framework. This framework is made of interactive combinations of brain regions roped together by two distinct functions: subject arousal and fluid intelligence.

Within AIM, it is the nucleus accumbens that controls positive subjective arousal. (It also mediates pleasurable feelings and addictive behaviors.) The insula controls negative subjective arousal (and is involved in "gullibility" in older populations, as well as feelings of disgust in all populations). As noted, parts of this system erode with age. In younger people, the insula is very active under conditions of negative subjective arousal. In older people, it is silent.

New learning is also affected. When seniors are given tasks that require them to make decisions based on recently learned information, they don't do very well. The more simultaneous the inputs, the worse it gets. The AIM network is in play here as well: it activates specific neurons in the prefrontal cortex (PFC) and temporal lobe to control fluid intelligence and decision making. As you age, however, the PFC—which normally talks to just about any brain region that will listen to it—quits interacting with the nucleus accumbens. This shunning affects certain tasks: ones where the brain needs to process new information and use it to update older, already processed information. You can also blame a failing working memory, which involves the PFC, too, and which illustrates you can never be too complicated when discussing brain circuitry.

Does this mean older people shouldn't be involved in decision making? Hardly. When tasks require information that was learned a long time ago (utilizing crystallized intelligence skills), seniors do just as well as their younger cohorts.

I direct your attention to one of the early scenes from Steven Spielberg's 1977 classic, *Close Encounters of the Third Kind.*

The scene begins in an air traffic control center, resounding with the calming clipped cadence of a gray-haired controller (who sounds remarkably like Morgan Freeman). He's seated in front of a radar screen, dealing with a hair-raising emergency. A number of commercial pilots are being buzzed by a UFO, and they're all worried about potential midair collisions. People gather around the older controller as tension builds, chattering excitedly, creating a noisy, confusing auditory environment. With hundreds of lives in the balance, an emergency ping suddenly sounds, warning of an imminent collision.

You'd think the older controller would be furious with his colleague's dangerous, all-at-once chatter. Or at least, distracted and nervous. Not this guy. He remains as calm as Quaaludes. The controller authoritatively issues a series of instructions, settling everyone down, and the crisis passes. Just before the scene ends, he asks this of one airplane: "TWA 517, do you want to report a UFO, over?" as if he were asking what the pilot had eaten for breakfast. The pilot declines.

What is going on in the mind of this extraordinary professional? How does he make these rapid-fire decisions? He seems to fly in the face of the data we just discussed, where simultaneous decision-making skills are increasingly compromised in older brains. Yet this isn't just Hollywood magic.

The controller saving the day wasn't some inexperienced wet-behind-the-radar whelp. He was an experienced professional, fortified with crystallized cognitive muscle. And no wonder. The job required him to keep his mind in the cerebral gym eight hours a day, exercising specific regions of the brain every time he showed up for work. Even though his mind might have been statistically deteriorating, his individual talent was better than anyone else's in the room. This is how nurture interacts with nature.

Brain games

You don't have to sit sphinxlike in front of a radar screen all day to reap the cognitive benefit of experience. The research is increasingly clear that you can exercise your attentional states at home. You'll still need a screen all right, but you won't need an airport. You'll just need some video games.

Yep, you read that right. Video games. For seniors. Especially BTPs, short for brain training programs.

A few years ago, you wouldn't catch me dead writing such a sentence, and with good reason. Have you heard of the company Lumos Labs and its suite of brain-training programs under the rubric Lumosity? Years ago, the company claimed that if you played their game-based BTPs for just a few minutes a day, you could ward off the most feared cognitive boogeymen of the over-sixty-five crowd.

These included memory loss, dementia, and even Alzheimer's. Close inspection showed the games had no such effects. The Federal Trade Commission sunk its canines into the company, initially fining it $50 million (later reduced to $2 million) for misleading the public. The FTC also ordered Lumos Labs to provide existing customers a quick financial way out. It was part of an "it's about time" crackdown on brain-training programs. *Jungle Rangers* (claiming to reduce ADHD symptoms) and *LearningRX* (claiming to treat severe cognitive impairment) came under similarly expensive scrutiny.

Shoddy research extolling the benefits of brain training still abounded with the frequency of winter flu. But other studies showed promise. Soon, responsible scientists gathered on both sides of the argument (contradictory voices are a terrific indicator of robust engagement, always a hopeful sign in science). Consider these two groups of scientists. The year before the FTC's Lumos complaint was filed, the first group (more than seventy scientists strong) signed a petition saying BTP was "baloney." Quote: "We object to the claim that brain games offer consumers a scientifically grounded avenue to reduce or reverse cognitive decline when there is no compelling scientific evidence to date that they do."

A contrarian chorus of researchers (about 120 strong), led by famed neuroscientist Mike Merzenich, provided the opposite voice: "No one is claiming that brain games will transform an average Joe into a Shakespeare or an Einstein. But there is plenty of evidence that computer-based cognitive training offers real benefits for certain populations. Most notably, it can cut an older person's risk of having a car accident in half."

These researchers faulted the skeptics for being not only hasty but also ignorant. Exhibit A was a pile of research papers showing that if you designed the games well and designed the evaluation instruments even better, you could check your doubts at the door. Hundreds of studies demonstrated cognitive benefits, they said. Though most agreed with the specific FTC complaints, they argued that ignoring the young science of cognitive training simply because it was young was, well, juvenile.

Today, with increased publication of higher-quality studies, the data show clear trend lines, and most are positive. Such is the deceptive charm of science, which builds consensus slowly, demands lots of arguments and hurt feelings, and continually inflates and deflates egos. Some programs need further work, and all could use more rounds of replication, but youth shows real signs of maturing. Lumos Labs matured, too: it now describes itself as "on a mission to advance understanding of human cognition" and talks about additional research. In the next pages, I'll describe a few brain-training games that have survived the withering fusillades of peer review, coming out bloody but unbowed.

Speed demons

I remember my first experience with a video game the way some people remember their first love. The game was called *Pong*. The machine was in a bowling alley, embedded in a yellow stand that looked like a snail's eyestalk. *Pong* was a simple electronic version of table tennis but, man, was I hooked! I eventually graduated to more complex gaming experiences (my next love-jones was with the game *Myst*). I tell you this to admit to my heaping serving of confirmation bias when it comes to advocating for video games. Fortunately, when talking about BTPs, my advocacy has a lot of independent empirical support.

Brain trainings to this day are as simple as *Pong*, and for good scientific reasons: less complexity means fewer uncontrolled variables. You get cleaner numbers and clearer findings. The best ones measure something researchers call "far transfer" effects. Many less-well-designed brain-training exercises (and that's most of them) improve only one thing: your ability to do well on the brain-training exercises. This unsurprising result is called "near transfer." What you really want is bleed-through, the ability to play a game and have it affect an unrelated cognitive process (perhaps changing processing speed or improving your memory). That's the definition of the far-transfer effect.

I am pleased to report that playing a few simple, lab-designed games has powerful far-transfer effects on cognition, as long as you play them the way the researchers intended. Let me describe one well-designed study employing a very simple speed-of-processing game: Imagine you are in front of a computer screen when two images flash suddenly, and briefly, into view—one in the center, one on the side. Your job is to answer questions about the experience. What object was in the center? What was on the side? Where on the screen did the peripheral image show up? In the true spirit of gaming, the better you get at answering these questions, the harder the game becomes. The images appear on the screen even more briefly. Pesky distracting images show up. Your speed and accuracy are measured throughout.

A group of researchers from Johns Hopkins and the New England Research Institutes were interested in the effects of this training not only on processing speed but on possible effects on the chances of coming down with dementia, which is about as far a transfer effect as you can get. The researchers gathered a cohort of cognitively healthy seniors, average age seventy-four years. This was christened the ACTIVE (Advanced Cognitive Training for Independent and Vital Elderly) study. The cohort was randomly assigned into four groups. One group did nothing (the control), one got training to improve memory, and one got training to improve reasoning. The fourth group was exposed to the processing-speed game for ten sessions, each about an hour long, over five or six weeks. (A random sampling also got "booster" exposures around one year and three years later.) The researchers then sat back for ten years, waited for the cohort to reach their mideighties, and looked for signs of dementia.

The results were a bombshell. At the end of ten years, people in the processing-speed group were 48 percent less likely to get dementia than any other group. That's astonishing. For one thing, the subjects were exposed for *less than a day's worth* of total training, yet the effects echoed with a cognitive sonic boom ten years later. That's what I call far transfer. For another, subjects in the group who'd received training to boost memory showed no improvement in those skills, essentially a waste of time. This puts into stark relief the strength of the positive findings.

This result has yet to be replicated, but it's still amazing. And it was not the first time researchers noted far-transfer improvements. A few years earlier, research led by the Mayo Clinic explored an auditory version of this same speed-processing experiment. Instead of two visual objects, the subjects in this study were asked to discriminate between two sounds played one after the other. The sounds might be two different pitches or two similar-sounding words ("sip" and "slip," for example). As the seniors improved on the test, the delay between the sounds got shorter and shorter. The seniors did this for an hour a day, five days a week, for eight weeks.

Similarly powerful far-transfer effects were observed: gains in processing speed led to memory gains. In terms of processing speed, seniors exposed to the training responded *twice as fast* as controls who received no training. Then Dr. Glenn Smith tested their working memory with the RBANS (Repeatable Battery for the Assessment of Neuropsychological Status). She said, "We found that improvement in these skills was significantly greater in the experimental group—about double."

Another audio game called *Beep Seeker*, developed at UC–San Francisco, improves working memory as well. You memorize a target tone, then hear a sequence of tones. Whenever you hear your target tone, you indicate it. This is harder than it sounds—made worse because as you get better, you hear more distracting tones, ones that increasingly sound like your target.

Researchers who use *Beep Seeker* aren't interested in tonal recognition, obviously. They're interested in distractibility, focus, and far effects. Could this training improve seemingly unrelated cognitive processes, like attention in other domains? Maybe working memory? The happy answer is yes and *yes*.

In one test of working memory, subjects scored a positive 0.75 (which is good), whereas untrained controls clocked in at a −0.25 (which is bad). Identical experiments were done with lab animals. The creatures showed the same far-transfer benefits.

Does that mean you should start playing those video games exactly as the researchers prescribed? That's exactly what it means. The game Smith used, developed by Posit Science, is commercially available. Others are sure to follow. You can find more details in the references at www.brainrules.net.

From arcade to prefrontal cortex

In preparation for writing this chapter, I happily played an arcade video game popular in my youth: an online variant of Atari's *Night Driver* (oh, the struggles researchers burden themselves with for science!). The game was still compelling after all these years, mostly because of its simplicity. You stare at a black screen, steering wheel in hand, and very quickly a "highway" appears. Your job is to navigate its various twists and turns. There isn't any highway, of course, or even pictures of one. There are only moving "roadside reflectors" flanking the screen's sides, small white rectangles fooling you into thinking you're cruising down a highway at night. Your job is to stay between reflectors, which whiz by you faster and faster as the game progresses. The best part? One video game reminiscent of *Night Driver* has now been shown in the lab to slow cognitive decline.

As reported in the journal *Nature*, UC–San Francisco scientists developed a game called *NeuroRacer*, which is like a three-dimensional daylight version of *Night Driver*. Subjects drive a virtual car through a landscape. They're warned that signs of various sizes and shape will suddenly pop into view while driving. To the probable delight of their grandkids, they're told to shoot down some of them—only ones of a certain size and shape.

Before playing, subjects in the study received a battery of cognitive tests measuring attentional states (like task switching) and working memory. They also were hooked up to an EEG (electroencephalogram) device. EEGs measure brain electrical activity in response to external stimuli. Researchers focused on activity in the prefrontal cortex.

Groups of older individuals (average age seventy-three) were then turned loose on the game, playing for a delightful four weeks. Brain activity was constantly monitored, and after one month, cognition reassessed. Untrained twenty-year-olds served as controls.

The results were a stunner.

First up were tasty far-transfer findings. Brain activity had shifted, especially in the prefrontal cortex, to a much "younger" pattern, as if the organ had been lifting weights in some mental gym. Pre- and post-behavioral assays confirmed the strengthening. Scores on a test of "working memory with distractions" improved dramatically with *NeuroRacer* (+100 in the video game group, compared with −100 in the no-game controls). Similar results were obtained in assays of "working memory without distractions" and the Test of Variables of Attention (TOVA).

Another finding concerned the stability of the boost, and this is the real headliner. *The improvements were still observable six months later*. When seniors who hadn't touched the game in half a year were measured against twenty-year-olds, they beat them! Here's a quote from the *Nature* paper: "[These findings] provide the first evidence, to our knowledge, of how a custom-designed video game can be used to assess cognitive abilities across the life span, evaluate underlying neural mechanisms, and serve as a powerful tool for cognitive enhancement."

Adam Gazzaley, leader of Team NeuroRacer, has enthused that his lab's creation may become "the world's first prescribed video game." That would be extraordinary, because we've known for years that attentional abilities decline with age. The preponderance of data, under the cheery thrall of the arcade, suggests it doesn't have to. And we owe it to a technology that began with an electronic paddle in your hand and ended with an electrode on your scalp.

To be sure, not everyone greets these with findings with a standing ovation. Critiques have ranged from sample size (the number of people studied) to real-world relevance (Does this help you remember you went to the pantry to retrieve juice?). The complaints are valid, though hardly deal killers. They fall under the sway of a scientist's go-to admonition: more research needs to be done.

You might recall from the introduction David Attenborough describing the way many smaller tributaries contribute to creating a large, smoothly flowing Amazon. If we think of that river like our brain's attentional states, the tributaries contributing to the flow include things we've covered: more friends, less stress, and learning the size of libraries. If you ask me, video games may constitute one of the most delightful contributing streams. As we'll see, they are hardly the only ones.

SUMMARY

Train your brain with video games

- Processing speed, the speed at which your brain takes in, processes, and reacts to outside stimuli, drops in the aging process. It is the greatest predictor of cognitive decline.
- Switching tasks becomes more difficult as you age. Consequently, it is easier to become distracted as you grow older.
- Specially designed video games like *NeuroRacer* have been shown to improve seniors' working-memory-with-distractions, working-memory-without-distractions, and Tests of Variables of Attention, beating twenty-year-olds who hadn't played the game.

your mind: alzheimer's

brain rule

Look for 10 signs before asking, "Do I have Alzheimer's?"

Soon there will be two kinds of people in the world.
Persons that have Alzheimer's and persons
that know someone that has Alzheimer's.
—Attributed to Dr. Mehmet Oz

We'll be friends until we're old and senile.
Then we'll be new friends.
—Anonymous

AUGUSTE DETER WAS CLEARLY troubled. At night, she dragged her bedsheets around the mental institution where she spent her last years, screaming at no one for hours on end. Though she was a frail woman, she could be physically assaultive, a danger to those around her. She was also mentally confused, emotionally disorganized. One interview, recorded with (and by) her doctor, began like this: "What is your name?" the clinician asked. "Auguste" was the reply. "What is your husband's name?" She hesitated a second. "Auguste, I think." "Your husband?" the physician repeated. "Ah, my husband!" she repeated, not understanding the question. The doctor continued, "Where do you live?" This question surprised her. "Oh, you have been to our place!" she exclaimed. "Are you married?" the doctor asked. Deter was hesitant, blurting out, "Oh, I am so confused." She sensed something was amiss, at one point declaring, "You must not think badly of me." The physician continued probing: "Where are you at the moment?" She responded rather incoherently, as if hearing another question: "We will live there" is all she said.

Deter was actually in Frankfurt, Germany, interred as a mentally ill patient at the psychiatric facility. The interviewer, however, was no ordinary doctor. His name was Dr. Alois Alzheimer. He was taking notes on the very first person ever diagnosed with the disease that would eventually bear his name.

Auguste Deter died in 1906, and Alzheimer was allowed to examine her brain in detail, one piece at a time. He found what has become Alzheimer's famous cellular signature, the odd fibrils and even odder plaques marbling Deter's brain like the fat of a rib eye steak. This damage was invoked to explain her mental condition, at the time called presenile dementia.

It's a condition that causes terror to this day. "Am I getting Alzheimer's?" is one of the most anxious questions any senior can ask. Your brain turns into your own personal gestapo, questioning every slip of the tongue, interrogating every lost cell phone experience, feeling tortured each time a familiar person's name is forgotten. The question drives patients, clinicians, and researchers alike crazy. That's because the answers are so unclear. Teasing out typical everyday aging from abnormal brain pathology is one of the greatest challenges the field faces, made worse because it's already among the greatest concerns of aging patients.

This chapter is all about what we currently know about Alzheimer's, how to detect it, how to differentiate it from mild cognitive impairment, and what we've learned from an extraordinary study of nuns. Yes, *nuns*. I will warn you there aren't many raindrops on roses in these next pages. At this point, we are still trying to define exactly what dementias like Alzheimer's actually are. For most researchers, this slow progress is definitely not one of their favorite things.

Mild cognitive impairment

There's a twilight zone between typical functioning and the beginnings of something worrisome. Clinicians use the term "mild cognitive impairment" (MCI) to describe it. It's almost always cumulative, the dysfunction starting imperceptibly, then gaining steam. Or not. We have no test that a clinician can use to determine what advice to give. That's because many types of MCI exist, and we are just now learning to differentiate them from one another. Research shows the brains of some people who died with MCI (note I did not say died *of* MCI) actually had thousands of tiny pinprick leaks in the blood vessels of their brain. Think mini-strokes. Others had a pre-Alzheimer's-like condition, showing beginning accumulations of classic clumpy plaques. Others have what looks like a pre-Parkinson's dementia or a pre–Lewy body dementia or a pre-*nothing*. (We'll talk about these various pathologies shortly.) The brains of some people with obvious mild cognitive impairment look perfectly healthy at autopsy, no frank physical signs at all.

What should we do? Current estimates suggest that between 10 percent and 20 percent of all people over the age of sixty-five already have mild cognitive impairment. So let's start there, then work our way toward Alzheimer's. What are the behavioral symptoms that suggest your brain may be listing away from typical aging and taking on pathological water? Most clinics provide a list of behaviors to watch out for; one of the best comes from the Mayo Clinic. They divide the "what to watch out for" into two familiar categories:

Cognitions

You forget your car keys. You forget appointments. You lose your train of thought, often. These changes in memory are called amnestic MCI. You might find it increasingly difficult to navigate familiar terrain. You become overwhelmed with even simple decision making. You misjudge the sequence of events necessary to complete a task or the time it takes you, or both. These changes are called nonamnestic MCI.

Emotions

Your behavior becomes increasingly socially "inappropriate." You are more impetuous, more reckless, and show increasingly poor judgment. These symptoms can be accompanied by mental health issues, like depression and anxiety.

How do these differ from all the aging deficits we've discussed so far? The truth is they don't. The one cardinal differentiator may come from an item on Mayo's list: *your friends and loved ones begin to notice something is wrong.* They observe you still performing all of life's daily tasks (that's what ties the diagnosis to MCI rather than dementia), but you're clearly struggling in one or more areas. You may successfully hide your disability for a time, fooling even the most insightful loved one. But if the condition worsens, the false front may crumble. At the point where the cognitive fissures become visible to someone besides you, action may be warranted.

What should you do? If you have some of these symptoms, or a loved one has some of these symptoms, a medical evaluation from the family doc is a good place to start. Most clinics begin with an assessment of mental status and/or mood ("affect"), perhaps along with a neurological examination, testing things like reflexes, balance, and various sensory abilities. Almost always, the physician recommends embracing lifestyle behaviors related to preventing strokes.

But here's the rub, and some partial good news: some people never progress beyond the symptoms mentioned. They live a long, happy life with MCI. Reinforcing a delightful staple of English fiction, they simply become the eccentric aunt or uncle. Others, of course, have mild cognitive impairment for a while, then get noticeably worse and begin displaying other symptoms. At the point where daily function is impaired, they are leaving MCI in the rearview mirror, traveling toward dementia. Thus you can think of MCI as a prophet who may be predicting a gathering storm. Or, rather, one of several potential storms.

Robin Williams

I'd been laughing ferociously with comedian Robin Williams since college, marveling that even his voice acting could put me in stitches ("You ain't never had a friend like me!"). I wasn't alone. You could feel the audience anticipation rise to DEFCON levels when he appeared on talk shows. Williams's comedic mind was always ready to detonate like a nuclear explosion. It's been a long time since he died, but his death still feels like an open wound.

Williams was diagnosed with Parkinson's disease several months before his death by suicide. The autopsy revealed something else, too. Williams had diffuse Lewy body dementia, a type of illness for which mild cognitive impairment can serve as a gateway.

Yes, the big thug in the room is Alzheimer's disease, responsible for up to 80 percent of age-related dementias. But it's not the only dementia out there. I want to describe a trio, beginning with the one that felled Williams.

Dementia with Lewy bodies

Williams's diagnosis wasn't uncommon. Lewy body dementia is the second-leading cause of dementia in the United States, accounting for between 15 percent and 35 percent of all dementias, depending on the study. It's named after the German scientist Frederic (Fritz) Lewy, who first noticed tiny dark dots around the neurons of people who had died from "senility." We now know those clumps are abnormal knots of the protein alpha-synuclein. The symptoms they cause include sleep disturbances, motor imbalances, memory losses, visual hallucinations, and then Alzheimer's-like behavior. We don't know why the knots cause dementia; we don't know how to treat it; we don't even know how people get it. In recognition of our ignorance, we call the disease's origin "idiopathic," a term over which Robin Williams probably would have cracked up.

Parkinson's disease

The second dementia is one not famous for being a dementia at all. Parkinson's disease is most notorious for causing people to lose motor control—arms flailing, legs refusing to follow gaiting instructions. Famous sufferers include Michael J. Fox, Muhammad Ali, and Billy

Graham. It's named for James Parkinson, a nineteenth-century British physician, who originally called it "Shaking Palsy."

That was a good name but also a tad incomplete. Although Parkinson's is a movement disorder, later stages almost always include dementia, cognitive disorders like changes in ability to focus, or affective disorders like depression and anxiety. Parkinson's disease occurs when brain cells in specific regions start dying off, like those in the substantia nigra (in the lower middle of your brain). No one knows why this cellular genocide occurs, though it may be related to a familiar villain—alpha-synuclein. Indeed, people with Parkinson's often have Lewy-like bodies hanging around their dying nerves.

Frontotemporal dementia

The third disease comes early. Frontotemporal dementia typically strikes younger people (around age sixty, though it can even hit twenty-year-olds). Language deficits are a symptom, but the biggie is a striking change in personality. You see wildly inappropriate behavior, such as shouting at strangers, hitting people, gorging on food, and exhibiting a marked indifference to loved ones. Frontotemporal dementia also can include repetitive behaviors, such as talking about the same subject over and over again, continually mowing the lawn, or walking the same path repetitively. It is neurodegenerative, with progressive damage to the frontal lobes (the regions behind your forehead) and temporal lobes (the ones next to your ears). No one knows why it occurs.

Then you have the vascular dementias, which cause cognitive mayhem the same way strokes do, by leaking small amounts of blood into the brain. There's Huntington's disease, the same dementia that claimed Woody Guthrie. There's even one that may be communicable, Creutzfeldt-Jakob disease, mediated by a particle called a prion. Fortunately, it is among the rarest of the group.

Unlike the big boy on the block. On both financial and humanitarian grounds, Alzheimer's may be one of the costliest diseases ever to strike the modern world. It's time we discuss it in detail.

Alzheimer's disease: an overview

Alzheimer was really on to something with his patient Auguste, of course, though exactly what he thought was wrong was a matter of conjecture. That's not unusual; at one time or another, virtually everything about Alzheimer's disease has been subject to debate and speculation. Even Dr. Alzheimer's original findings were considered suspect after his death. Fortunately, he kept careful notes—and his brain-tissue slides. Modern scientists reexamined his work and confirmed it.

Though the science behind the disease remains controversial, its economic blast radius is not. Whether measuring human capital or financial treasure, Alzheimer's costs the planet a bundle. Dementias of any kind rank No. 5 on the list of the biggest causes of death in the developed world, but it ranks No. 1 in expense. That's because a patient can live for many costly years after diagnosis (a decade between diagnosis and death is not uncommon). In the United States alone, where 5.4 million people were afflicted with the disease in 2016, the cost of their care was $236 billion.

These numbers might not give society such economic acid reflux if the research world knew exactly what it was studying. You might be surprised to learn that it doesn't. Dr. Alzheimer's slides showed clearly that Deter had brain damage. Yet further research has shown clearly that not all patients diagnosed with her behavior also exhibit her brain pathology. More puzzling, not all patients exhibiting Deter's brain pathology exhibit her behavior. This field is currently mired in contradiction, especially at the molecular level.

By far, the leading theory for Alzheimer's origins is something called the amyloid hypothesis, which we'll dive into in a bit. Not every researcher thinks it is adequate as the sole explanation for all observed pathologies. Or even a partial explanation. Some researchers (I'm one of them) believe it should be called Alzheimer's *diseases*—for there's almost assuredly more than one type. Partially because of this ambiguity, no one test can definitively detect Alzheimer's. If you visit your doctor's office worried about Alzheimer's, you'll undergo the same tests issued for any form of dementia. Only when certain behaviors are ruled out might your doctor say, "You may have Alzheimer's." And that's exactly how doctors couch it, for an important reason:

They don't really know if you have Alzheimer's. Nobody does. Even an autopsy is not necessarily definitive, for reasons we'll get into shortly.

It's important to visit your physician, however, as soon as symptoms actively interfere with your ability to function on a daily basis. It's one thing to go fetch something from the basement and forget what you came for. It's another thing to go down to the basement and forget where you are.

Alzheimer's disease: warning signs

Excellent checklists have been developed over the years to help loved ones determine whether a person has Alzheimer's or is simply guilty of being a senior. One of the best is the Alzheimer's Association's "10 Warning Signs of Alzheimer's Disease," which I'll summarize here. The ten signs can be organized by topic: memory, executive function, emotions, and general processing.

Memory

The first four signs, unsurprisingly, involve memory:

1. Memory loss that disrupts daily life

Working memory naturally erodes with age. When loved ones routinely forget important dates and appointments, however, or abnormally rely on physical prompting strategies (like Post-it notes), it's time to see someone. Ditto if they begin requiring information to be repeated over and over again.

It's a frequency argument. You don't need to worry if they occasionally forget appointments or someone's name. You do need to worry if it happens all the time.

2. Difficulty completing familiar tasks

If loved ones forget how to balance checkbooks, which way to drive to the store, or the rules of a beloved board game, concern is warranted. As Alzheimer's tightens its grip, people have increasing difficulty completing familiar routines. Thus it's okay if they forget that Monopoly was invented by Parker Brothers. It's not okay if they forget how to play it.

3. New problems with words in speaking or writing

As discussed, core language abilities seldom erode with age. So take note if loved ones start tripping over their own words, have progressive difficulty following conversations, or routinely stop mid-sentence because they suddenly don't remember how to continue. It's typical aging not to find appropriate words. It is atypical aging not to find any words at all. The same difficulties occur, interestingly, with written communication.

4. Misplacing things and losing the ability to retrace steps

One unusual feature of Alzheimer's is the inability to re-sequence information, such as attempting to retrace one's steps when searching for misplaced objects. It's problematic because people with early Alzheimer's routinely put things in odd places (perfume in a freezer, medicines in a soap dish). Misplacing things happens all the time. But putting Chanel where it doesn't belong is worrisome.

Executive function

Executive function naturally erodes with time, but precipitous life-interfering changes such as these aren't natural at all:

5. Challenges in planning or problem solving

An increasing inability to follow a plan (like a recipe) or devise one (like making budgetary room for an expense) is a red flag. So is an increasing loss of concentration, causing seniors to take increasing amounts of time to perform regular tasks like paying monthly bills. Forgetting to write a check to the cable company at month's end is not necessarily cause for alarm. Forgetting to write checks at all is.

6. Decreased or poor judgment

Executive function involves decision-making skills, which aberrantly erode with Alzheimer's. Deficits show up in everything from making poor financial decisions to forgetting to brush one's teeth. You'll often see other changes in grooming habits. It's normal for loved ones to occasionally forget where their glasses are. It's not normal for them to forget to put their pants on. Or to give away their retirement savings to the next homeless man they see.

Emotional processing

This next pair of warning signs involves changes in mood and emotional regulation:

7. Withdrawal from work or social activities

An early sign of Alzheimer's may be social secession—a withdrawal from familiar and formerly pleasurable social activities. As discussed in the opening chapter, such withdrawals can have dramatic negative cognitive effects, made all the worse if a pathology like Alzheimer's is involved. Often a person is well aware of all the deficits occurring and, ashamed to tell anybody about them, withdraws.

8. Changes in mood and personality

Another early sign of Alzheimer's may be related to mood changes. People with Alzheimer's may become paranoid, anxious, fearful, or increasingly emotionally disorganized. They might react inappropriately to the normal ups and downs of life, particularly when not in familiar surroundings. While it's typical for seniors to both develop and rely on reassuring daily habits, it is not typical to become catastrophically upset when such routines are disrupted.

General processing

The last two warning signs involve processing issues not explicitly related to memory, executive function, or emotional regulation:

9. Trouble understanding visual images and spatial relationships

Living with experienced eyes means some wear and tear: older people can't see as well. But with Alzheimer's, the loss is not of visual ability but of visual perception. People lose the ability to judge distance, understand color or contrast, and interpret the spatial relationships between objects. This naturally affects the ability to drive.

10. Confusion with time or place

You're probably most familiar with this one. Seniors losing track of time or where they are is a hallmark of Alzheimer's. They increasingly focus only on the immediate world, which may be related to a faltering ability to plan. Their internal GPS begins flickering. Wandering, accompanied with bewilderment, fear, and anger about where they end up, becomes a huge problem in later stages. It's normal to momentarily forget the day of the week, or even transiently to forget where you are when walking around your neighborhood. It's not normal to wander the neighborhood at midnight, wondering how you got there, yelling at the top of your lungs at no one in particular.

For those of you struggling with the dilemmas of a loved one who has the disease, I could not recommend more heartily the information found on the Alzheimer's Association website, www.alz.org.

Lessons from a president

Two letters penned by the late president Ronald Reagan stick in my memory. The first was addressed to my mother, Doris Medina, who was (briefly) a rising starlet in Hollywood in the late 1940s. She prudently joined the Screen Actors Guild, then headed by the actor Ronald Reagan, and quickly received a letter from him. It was surprisingly personable, welcoming her both to Southern California and SAG, signed by himself and his then wife Jane Wyman, with a scrawl from daughter Maureen.

The second letter, written in 1994, wasn't addressed to my mother but to the world. Reagan was announcing the way he was going to die.

> *I have recently been told that I am one of the millions of Americans who will be afflicted with Alzheimer's disease. . . . Unfortunately, as Alzheimer's disease progresses, the family often bears a heavy burden. I only wish there was some way I could spare Nancy from this painful experience. When the time comes, I am confident that with your help she will face it with faith and courage. . . .*
>
> *I now begin the journey that will lead me into the sunset of my life.*

I had many political differences with Ronald Reagan, just as I have with most politicians. But in this equally humanizing, humbling moment, there was no place for bickering. There was only a great, vulnerable old man, struggling with one of the most brutal ways to die. It made me cry.

President Reagan would not slip his mortal coil for another ten years. The average is four to eight years, which is why Alzheimer's is sometimes called the Long Goodbye. This is no ordinary aging, however. For people living with Alzheimer's at age seventy, about 60 percent will be dead before age eighty. For people without Alzheimer's, only 30 percent will be dead by age eighty. Thus Alzheimer's roughly doubles the risk of death. It's the sixth leading cause of death in the United States, regardless of age.

Every sixty-six seconds, someone develops the disease. That statement is a bit misleading, however, and for an unexpected reason. There's now strong evidence that the disease actually begins ten to fifteen years before observable symptoms surface. Some reports put the delay at twenty-five years. This means that by the time you forget how to drive to the mall, you've been living with Alzheimer's for more than a decade. So we should say Alzheimer's is *detected* in someone every minute or so. Currently, it afflicts about one in ten Americans over the age of sixty-five, more than five million people. As baby boomers age, that figure is expected to triple by 2050.

The disease progressively turns people's lives into architectural ruins in three overall stages: mild (wandering begins, personalities change), moderate (more memory losses and confusions, increased dependency on others), and severe (collapse, complete dependence on others). These categories are not set in stone, however, because Alzheimer's is very individually experienced. The progression inevitably, inescapably vectors from mild to death, but different people suffer different things along the way. Yet I do mean inevitably, inescapably. Says a pamphlet put out by the Alzheimer's Association, the same organization that gave us the warning signs: "Alzheimer's is the only cause of death among the top 10 that cannot be prevented, cured, or even slowed down."

That hasn't stopped researchers from setting their strong shoulders to the task of finding cures, of course. Progress has been slow, Sisyphean, and controversial but hardly nonexistent. It is to this progress, beginning with the gene work, that we now turn. We've already spent billions on the problem, and we are likely to spend billions more before we find something. The research fruit of some of those funds concerns DNA. There appears to be a genetic basis for some forms of the disorder (watch out if you are a woman possessing a gene variant called ApoE4). However, these heritable forms represent a minuscule 5 percent of all the known cases of Alzheimer's, according to Yale researcher Vince Marchesi. What's causing the other 95 percent? We don't really know.

Some think Alzheimer's may actually represent a cluster of diseases. Nonsense, say others, pointing to the Everest-size mountain of work filed under "Amyloid Hypothesis" as evidence. It's to this hypothesis that we turn next, starting our controversial tale in Manhattan, New York—with 1980s mobsters.

The amyloid hypothesis

It was the bloody mob hit of 1985. Paul Castellano, unpopular head of the Gambino family, was gunned down in the middle of Manhattan during rush hour, just as he stepped out of his car. The person arranging the hit didn't do the actual killing; we all know most of the mob's dirty deeds aren't done by the people who order them. Castellano's assassination was a bit unusual because the man who *did* order the contract, John Gotti, watched the killing from a car across the street.

This distance between mob bosses and their hit men has direct relevance to the amyloid hypothesis. The gangsters here are two sets of proteins: one ordering hits on aging neurons; the other carrying them out. To understand how this works, we have to know something about how cells make proteins.

As you know, the cell body of a neuron contains a nucleus, a round little ball brimming with command and control functions. Those responsibilities occur because of slender DNA molecules crammed into its salt-watery sphere. One way this tiny helical titan exerts its power is by producing instructions to make proteins, a class of molecules as critical to life as breathing. Making proteins, however, involves solving a little problem with big implications. Whereas DNA is locked tight in the nucleus, the protein-manufacturing sites are permanently locked out of headquarters, forcing them to reside in the cell body (the cytoplasm). Immobile DNA resolves this by making tiny strips of portable instructions, called messenger RNAs, which get smuggled out of the nucleus and into the cytoplasm. Once those arrive, molecular mechanisms read the message, send for the protein-manufacturing machinery, and go to work. New proteins soon roll off the assembly line in large, often ungainly, often *useless* forms. To make them functional, many undergo an editing process, irrelevant parts snipped away, important parts rearranged, small molecules added. It's called post-translational modification, which turns out to be important to the amyloid hypothesis.

Looking microscopically at the brains of some deceased Alzheimer's patients is like seeing the messy aftermath of a mob hit. There is the detritus of dead nerve cells, holes where there used to be healthy tissues, and strange flotsam called plaques and tangles. Plaques are clumps of amyloid protein, which look like big fuzzy meatballs lying outside surviving cells.

Amyloid normally undergoes post-translational modification after manufacture, but this editing process goes awry in Alzheimer's patients. The reasons are probably genetic. The dysfunction creates an accumulating pile of sticky fragments called A-beta (spelled Aβ). These assemble into toxic clumps and even deadlier soluble, semi-clumped aggregates. This is like forming the equivalent of an angry Mafia boss. The aberrant structures soon order the death of neurons. Though some do their own killing (synapses are a favorite target), they leave much of the dirty work to another protein. You can think of *that* protein as the hit man.

The trigger happy assassin involves these tangles. Looking like knots of deadly snakes, these structures assemble inside living neurons. They're made of proteins we call tau, which in their normal form are common and helpful. For reasons not well understood, amyloid Mafia bosses order neurons to make a modified, fibrous, lethal form of tau. It's that form that destroys the interior of neurons, killing the cells, which then releases them into the intercellular space, where they are free to continue killing other neurons. They create a path of destruction, from destroyed synapses to destroyed neurons, leaving a gory mess inside the brain. In the final stages, an Alzheimer's brain shrivels up like a dried sponge.

At least that's what some people think.

There are lots of reasons to scratch one's head about the amyloid hypothesis. The main reason is that some people get all the plaques and tangles, but none of the disease. Some get the full disease and none of the plaques and tangles. The first subjects to show us this? Nuns.

The Nun Study

"I only retire at night!" Sister Mary boldly declared to her colleagues, defiant as a teenager. And she meant it. Then in her mideighties, she was still a force to be reckoned with, all ninety pounds of her in a four-and-a-half-foot frame. Sister Mary taught junior high school for nearly seven decades. Even when "retired," she still held court with younger nuns, serving as the convent's forceful dynamo until her batteries ran out at age 101. Sister Mary was enrolled in the famous Nun Study, generously donating to science not only her biography but her brain.

The Nun Study was the brainchild of Dr. David Snowdon, a researcher who regularly studies the brains of patients with Alzheimer's after they've died. His problem, familiar to all such investigators, was finding enough relatively disease-free seniors willing to donate their brains after death, serving as vital controls. Bonus points for people unencumbered with interfering lifestyle confounders, like alcoholism or chronic drug use.

The solution turned out to be just a few miles south of him. There was a Roman Catholic convent not far from the University of Minnesota (his lab at the time), and he hit on an idea. Would the School Sisters of Notre Dame be willing to partner with him in a long-term research relationship? Many were getting along in years, some already exhibiting behaviors associated with Alzheimer's. The convent and its sisters could present an ideal research opportunity. Their life courses were well documented, mostly free of the previously mentioned encumbering lifestyle confounders. The idea was to measure their behaviors while alive, then have them donate their brains to Snowdon's lab at death. He could then study their neuroanatomy in greater detail.

The response of the Sisters was overwhelming (they were a teaching order, after all). Nearly 680 nuns enrolled, all over age seventy-five, and in 1986, one of the most valuable research efforts in the field, simply called the Nun Study, was born. With funding from the National Institute on Aging, researchers swarmed to the convent over the ensuing decades. They were armed with batteries of assays, including cognitive, physiological, and physical strength tests. When a sister died, her brain was donated to science and examined by the lab.

Sister Mary—the gold standard for successful cognitive aging, Snowdon once said—was next.

Given his observation, you'd think Sister Mary's autopsy would reveal a brain preserved with joyous functionality—understandably worn but still intact, maybe even youthful. That is exactly what he *didn't* find. Sister Mary's brain was a neuroanatomical mess. It was filled with the plaques and tangles and cellular pathologies associated not with gold standards, but with Alzheimer's. That she remained cognitively immune to its effects seemed miraculous.

To add spice to this mystery, Sister Mary isn't all that unusual. Researchers now know that 30 percent of all people with no signs of dementia have brains choking with the molecular detritus of Alzheimer's. About 25 percent of people who have Alzheimer's disease show no significant accumulations of plaque. The statistics appeared to breathe sulfur into the lungs of the amyloid hypothesis.

Pharmaceutical companies have attempted to treat Alzheimer's by targeting amyloid directly. One drug, awkwardly named solanezumab, has received special attention. It binds to that deadly Aβ protein fragment in the fluids surrounding the brain. This binding increases its elimination from the brain. The idea was that if you could lower the concentration of Aβ available for mayhem in deeper brain tissues, you would reduce damage.

It cost Eli Lilly almost $1 billion to discover that this idea was wrong. Solanezumab does nothing to reduce even mild dementia in Alzheimer's patients. Lilly abandoned testing in November 2016. One research paper had had the cheek to put in its title: "When There's No Amyloid, It's Not Alzheimer's." Now a critic trumpeted, "The amyloid hypothesis is dead."

In my view, writing a molecular epitaph for the idea is a bit premature. Even the stoutest critics believe amyloid plays *some* role in Alzheimer's. But if plaques and tangles aren't the whole story, what is? Are researchers even asking even the right questions? Some suggest they're not.

Such accusations are fueled in part because of comorbidity studies (comorbid means "found with"). Researchers have known for years that many patients who die of Alzheimer's also had other problems with their brains. For example, amyloid deposition often is concurrent—comorbid—with the presence of Lewy bodies. You recall that Lewy bodies are those tiny, dark round dots that filled the brain of Robin Williams. The offending dots are α-synuclein proteins. Their association with Aβ is not trivial. This mixed pathology is observed in more than half of all patients diagnosed with Alzheimer's. Could it be that the amyloid hypothesis should be rechristened the amyloid-and-α-synuclein hypothesis?

Another theory has more in common with scraped knees than black dots. Some researchers believe the presence of Aβ isn't what triggers Alzheimer's, but rather the presence of inflammation in the brain—unsurprisingly termed neuroinflammation. It is true that inflammation often precedes the formation of Aβ. In this view, the primary culprits are cytokines: molecules that induce brain-wide, even body-wide, irritations. These tiny irritants overstimulate the immune system of the human brain, driving damaging responses. This leads to the neurodegeneration (synapses make an especially ripe target) usually associated with Alzheimer's.

These ideas, compelling as they may be, are still shots in the dark. And that's kind of where we are with Alzheimer's. At this stage, we don't know how to cure it. We don't know how to slow it down. We don't really even know what it is. I told you this chapter would not be a lot of fun to read. But the Nun Study outlines a potentially powerful direction for Alzheimer's research. It involves no drugs or genes, but simply written autobiographies. I've saved this most intriguing result for last.

Predicting Alzheimer's in your twenties

The convent required that the nuns write their life stories when they joined. The women tended to be in their twenties then, and their writing samples were archived. That gave Snowdon an idea. When the Sisters died six decades later, he had their essays undergo neurolinguistic examination. Why? Snowdon now knew who acquired dementia (and amyloid) and who didn't. This allowed him to ask an interesting question: Could you predict who got Alzheimer's in their eighties simply by analyzing writing samples authored in their twenties? It's all correlational work for sure, which is why I used the word "potentially." But real research fruit was obtained.

The writing samples were analyzed for linguistic density, a complexity measure, and the number of ideas per sentence. Eighty percent of the nuns whose writings didn't meet specific neurolinguistic benchmarks—who scored low on linguistic ability—developed Alzheimer's. Only 10 percent of those who scored high on those same benchmarks did. Idea density was especially predictive.

What does that mean? Currently, nothing. Except that the damage associated with Alzheimer's may begin earlier than anyone has imagined, and it may be too late for treatment by the time dementia arrives. Perhaps the billion-dollar solanezumab really *does* work, confirming parts of the amyloid hypothesis, but the patients were too far gone to save.

These ideas point us toward the future of Alzheimer's research. And we have reasons to be cautiously excited. Researchers recently characterized a molecule that binds to amyloid plaques. It's called PiB, clumsily short for Pittsburgh Compound B. But instead of attempting to eliminate plaques, as solanezumab did, PiB causes the plaques to show up on PET scans. That's because PiB has been made radioactive. Scientists can now see how much plaque a person has accumulated in real time. That's valuable knowledge. Clinicians are able to look for potential amyloid Alzheimer's without waiting for an autopsy.

PiB is a valuable research tool, too. Because people of any age can be screened with it, researchers can follow patients over time, determining who's accumulating plaques and who isn't decades before dementia occurs. Such information would certainly be valuable for investigating the amyloid controversy. But it might also help with pharmaceutical cures. One joint research project actually under way, called the Alzheimer's Prevention Initiative, uses some of these ideas. As its name implies, it's a valiant attempt at preventing Alzheimer's, and it involves about three hundred members of an extended family in Antioquia, Colombia.

Many in this South American town carry what is arguably the deadliest genetic Alzheimer's mutation in the world. It's called PSEN1 (presenilin 1), whose gene product does that amyloid editing we discussed earlier. This mutation is particularly cruel. First, if you have it, you are 100 percent guaranteed to get Alzheimer's disease. Second, the Alzheimer's you acquire is a rare form called early-onset, with symptoms observable by your midforties. It still takes half a decade to die from, like most Alzheimer's, but it strikes you at the prime of life. The town has the highest concentration of this form of Alzheimer's in the world.

Researchers took the following three steps:

1. Screening

They flew younger, asymptomatic members of this town, in their midthirties, to a lab in Arizona. Some carried the gene; some did not. The lab screened each person's brain using PiB and the PET scan. Those with the gene had already started accumulating plaques.

2. Treatment

Some members received an antibody-based drug similar to solanezumab, ridiculously called crenezumab. In accordance with the gold standard for behavioral research (so-called double-blind studies where researchers don't know which subjects are treated), some people got the drug and others did not.

3. Waiting

Was the drug given early enough to ward off dementia? The researchers won't know for many years. (In a side experiment reminiscent of the nun's autobiographies, members of this Colombian family were given neurolinguistic evaluations. Sure enough, those people with the lethal mutation scored significantly lower in the analysis.) Even if this Alzheimer's Prevention Initiative is successful, it won't prevent all types of dementia. It won't even prevent all types of Alzheimer's. And there is still no cure for people in the throes of even its milder forms. But it does hint at something positive, which is important. Such lines of research are easily the brightest lights in this most shaded corner of geroscience.

Fortunately, for the many people who will never get Alzheimer's disease, there are other bright regions to explore in the world of the aging brain, and some real reasons to celebrate. We are now going to pop open the champagne and consider behaviors that can slow the aging process considerably. While it is currently not possible to arrest the aging process, there's a great deal we can do to make the experience more comfortable than could any generation before us. In some cases, we may even be able to reverse some of its effects.

SUMMARY

Look for 10 signs before asking, "Do I have Alzheimer's?"

- Neuroscientists have a tough job teasing out typical, everyday aging from abnormal brain pathology. Just because you might show symptoms doesn't mean a pathology exists.
- Mild cognitive impairment is the term clinicians use to designate the beginning of brain pathologies. MCI doesn't mean seniors are necessarily on the path to dementia, Parkinson's, or Alzheimer's disease. Many seniors live long, happy lives with MCI.
- Dementia is a catchall term for a cluster of symptoms related to a loss of mental function. There are many age-related types.
- One in ten Americans over sixty-five lives with Alzheimer's. It is the most expensive disease to treat in the world. The average amount of time people live with an Alzheimer's diagnosis before they die is four to eight years.

BODY AND BRAIN

your food and exercise

brain rule

MIND your meals and get moving

Those who think they have not time for bodily exercise
will sooner or later have to find time for illness.
—Edward Stanley (Earl of Derby), 1873

Life expectancy would grow by leaps and bounds
if green vegetables smelled as good as bacon.
—Doug Larson, newspaper columnist

PATTY GILL RIS, EIGHTY-SEVEN, was eating her favorite meal at the Hyde Park senior facility in New York when she began choking on a piece of meat; it was suddenly and lethally lodged in her windpipe. Her dinner companion saw immediately what was wrong and, alert as a mousetrap, sprang into action. He turned the victim around, thrust his arms under her armpits, placed one fist below her rib cage but above her belly button, and pumped upward three times. He was obviously doing a classic Heimlich maneuver. Out flew the offending protein. But only part. Two more times he performed this legendary medical move to get all the meat out.

The age of Ris's award-winning rescuer? Ninety-six years. His identity? Famed thoracic surgeon Dr. Henry Heimlich. Yep, *that* Dr. Heimlich.

Why do I bring up this interesting coincidence in a chapter on aging, exercise, and food? It's less to comment on Ris's choice of meal than to point out why Heimlich was able to save her. Doing the Heimlich at any age is physically demanding. But doing it at ninety-six years of age—thrice—is almost science fiction. Perry Gaines, maître d' at Hyde, said: "At his age, that's a very physical type of activity. To see him do it is a fascinating thing." Another employee confirmed that Heimlich, a resident of the senior facility for about six years, "is very active for his age. He swims and exercises regularly."

Heimlich was obviously in shape, a fact that would become quite clear if you viewed an interview with him. (He looked like an elderly James Taylor.) But that's not the only thing that would catch your eye. There was a light in Heimlich's countenance, coupled with a gentle attentiveness that is almost startling. His mind appeared as alert as his body. He was deliberate, observant, and possessed with an air of quiet decisiveness. You can see how he could have spent a lifetime successfully navigating tense surgical suites. And you could see how he'd still be able save a life at an age when most people are dead. Though he'd long since retired by the time he rescued Ris, his mind did not get the memo. He died in 2016.

These two ideas, mental attentiveness and physical exercise, run through this chapter like the marbling in Ris's piece of beef. We start with a fact that is, if you'll forgive me, a bit hard to swallow: mental attentiveness naturally erodes over time. But we won't dwell on this decline for long. There are powerful ways to elevate brain functioning that have to do partly with exercise and partly with food. Both were exemplified in the lifestyle of a famous ninety-six-year-old physician who gave more than one person a new lease on life.

Calm, cool, collected

We begin, like a Calvinist sermon, with the tough stuff. We're going to spend most of our time here together on a certain category of mental attentiveness: a complex suite of behaviors in your brain termed executive function. I've mentioned executive function several times in this book. Each time, I've promised to elaborate on the cognitive gadget later. That later is now. I start with one of the clearest expressions of executive function I've seen in my lifetime.

I remember vividly the day Osama bin Laden was killed. Not because I was watching the news, but because I was watching snippets of the 2011 White House Correspondents' Dinner, which had occurred the night before. President Obama was on the lectern that evening, cracking jokes at the expense of his guests with ease, smiling, seemingly relaxed. He had a few zingers for Donald Trump, praising him for putting the matter of Obama's birth certificate to rest, saying that no one was happier than the future forty-fifth president because "he can finally get back to focusing on the issues that matter—like did we fake the moon landing? What really happened in Roswell? And where are Biggie and Tupac?"

No one could guess that the day before President Obama had authorized US Army Special Forces to execute Operation Neptune Spear, the secret operation that would kill Osama bin Laden. The strike took place on Sunday, the morning after the dinner. But there was no telltale clue that evening. I saw no "night before" tension on the president's face, no distracting fifty-mile stare, no fidgeting, no uncalled-for sweat. Even when host Seth Meyers cracked a joke about bin Laden, all you could see was Obama's broad smile and easy countenance. Yet he was about to kill the man the assembled might of the American military had been seeking for almost a decade. He looked as if he were watching a sitcom.

That, folks, in a nutshell, is executive function.

Loosely stated, executive function (EF) is the behavior that allows you to get tasks done—and to be calm and civil while doing them. It is vital in many aspects of life, including running the free world.

Many diverse cognitive processes make up executive function, and all scientists agree on which neural acreage belongs to Club EF. Researchers agree that executive behaviors can be subdivided into two simple fiefdoms: emotional regulation and cognitive control.

Emotional regulation includes impulse control, which incorporates the ability to delay gratification. You might want to eat that artery-petrifying cheeseburger in a sports bar, but you choose the healthy kale salad instead. Emotional regulation also involves emotional control: the capacity to edit your emotions in a socially appropriate manner (not laughing at funerals, for example). These two regulatory components often work together. You ache to punch your boss in the nose after he has given you a bad performance review. Proper emotional regulation, and perhaps threat of a lawsuit, ensures that you don't.

Cognitive control is a flowing conduit of good sense. Its hallmarks include the ability to plan (creating steps in the pursuit of some goal), to adapt flexibly to changing circumstances, and to organize seemingly disparate inputs into manageable, organized rubrics. Add to that the capacity to shift attentional focus from one task to another, prioritizing inputs while avoiding distractions. Another card-carrying member of Club EF is working memory. That's our temporary storage feature that used to be called short-term memory (remember, from our memory chapter, Pixar's Dory?).

Given its importance to human cognition, you might expect that scientists have spent a long time looking at the neurobiology behind executive function. And they have. One of the clearest findings is that EF is developmentally regulated: specific, observable changes occur over time. Teenagers supposedly don't have much of it, for example—or they ignore the EF they do possess.

Remember being a teenager, or your kids being teenagers? Then you'll appreciate this snarky post floating about online: "TEENAGERS: Tired of being harassed by your stupid parents? ACT NOW!!! Move out, get a job, pay your bills . . . while you still know everything."

Teenagers, not surprisingly, have a different take on the dumb things they do. One mini manifesto online: "We're teenagers. We're still learning. . . . We cheat, we lie, we criticize, we fight over stupid things. We fall in love and end up getting hurt. We party till dawn, we drink till we pass out. . . . One day that's going to all pass. You can waste your time focusing on all the bad things, but one day you're gonna wish you were still a teenager. So make the most of what you have now, forget all the . . . drama and live your . . . life with a sexy smile on your face."

Just about everything in that quote is related to executive function: planning, decision making, navigating social relationships, preserving aspects of personality, maintaining self-control.

And the part of the brain responsible for all this is the prefrontal cortex, or PFC, that important bundle of nerves we talked about in chapter 3. The PFC is involved in nearly every aspect of executive function. That's not because it sits there isolated behind our foreheads being smart. It mediates executive function because it has befriended many other brain regions, reaching out via complex networks of neurons.

Vast systems of neurons connect the regions of the brain, as you know. They work something like interstate freeways, connecting one city with another. The PFC is a terrific example of a "city" with many neural highways connecting it to other regions. Technically, we would say the PFC has high "structural connectivity" to other regions.

Neuroscientists also think in terms of functional connectivity, which has to do with task rather than structure, and this occurs because the brain does not use all of its highways all the time. Some neural pathways are used in selective combination with others, connected to specific locations, to allow specific functions (hence the term). This is how the PFC mediates executive function.

These specific locations are familiar to you by now. The amygdala, working like a well-written romance novel, helps generate the experience of emotions. The neurological freeways connecting the PFC to the amygdala assist in the emotional regulatory components of executive function. Connections to the hippocampus, a region associated with long-term memory, assist in cognitive control. The PFC even has internal connections, as if the PFC befriended itself, involved in the formation of working memory.

There's dramatic growth in executive function during our toddler years, followed by a rest, followed by even more dramatic growth at puberty. Things don't actually settle down until we're in our midtwenties. And then, in old age, executive function begins to slide. To help me explain, here's a thought experiment regarding the city where I live.

Cracks, leaks, and potholes

I live in Seattle, Washington, a relatively small urban emerald (population 686,800) housing the world headquarters of a ridiculous number of corporations. From Amazon to Zillow, Nordstrom to Starbucks, many multinationals call Seattle home. Microsoft does, too, in a town just across the lake. And Boeing is *everywhere.*

Here's my thought experiment. These behemoths require massive numbers of people not only to run them but also to maintain and repair their infrastructure. What would happen to all of the companies' glittering progress if the repair and maintenance personnel in the greater Seattle area began to disappear? What would it look like to have things break and never get fixed?

When power failed, there would be no electricity. When pipes broke, no one would plug leaks, replace conduits, or mop up. Windows would remain broken, roofs would leak, building structures would eventually fall. The corporations would stagger, then fall to their urban knees. The roads connecting one giant to another would get pockmarked, crumble, and ultimately fail. It wouldn't take long for things to look positively postapocalyptic.

This type of erosion is exactly what happens to executive function. In our youth, the structures and connections get hobbled, but our repair mechanisms are active. Somewhere around age sixty, those maintenance mechanisms start retiring. "A man loves meat in his youth that he cannot endure in his age," Shakespeare once said. Normal wear and tear, increasingly, is not repaired.

Failure occurs at two levels. First, the highways that connect the PFC to those distant regions mediating executive function begin deteriorating. One study showed that 82 percent of executive function loss is directly attributed to a degeneration in the neural freeways the PFC used to stay in touch with its far-flung friends. Second, the brain structures linked together by those freeways—the cities in our metaphor—also begin to fail, buckling like abandoned towns. Research reveals that the hippocampus shrinks in an age-dependent manner. The PFC also loses volume.

These are critical losses. PFC neurons that support working memory do so by maintaining electrical activity through what are called excitatory networks. (This stimulation is sustained in the absence of any outside prompting.) When so many neurons are lost that one can observe structural shrinkage, maintaining internal network integrity becomes increasingly impossible.

So that's the bad news. We obviously need a helping or two of that good news discussed earlier. Look no further than iconic television producer Norman Lear. His life is an emblem showing us just how good the news can be.

Get your brain off the couch

For those of us watching sitcoms in the 1970s, Norman Lear was as constant a presence in our lives as oxygen. He was the driving force behind such shows as *All in the Family*, *Good Times*, *The Jeffersons*, and *Maude*. He never retired. In 2016, at the tender age of ninety-three, he embarked on a new TV show, a Latino reboot of another hit show of his, *One Day at a Time*.

His brain is still lightsaber sharp. In 2016 he appeared on the NPR quiz show *Wait Wait . . . Don't Tell Me!* Host Peter Sagal asked him, "So do you have any tips for those of us who would like to arrive at ninety-three as spry and as successful and happy as you are?" Lear responded, "What occurred to me first is two simple words. Maybe as simple as any two words in the English language: over and next. We don't pay enough attention to them. When something is over, it is over, and we are on to next. And if there were . . . a hammock in the middle, between over and next, that would be what is meant by living in the moment.

I live in this moment." Lear was really on to something neurological, though he may not have known it. Remember our discussion about mindfulness? Living in the moment is one of its signature attitudes.

The panel and host, usually quick to pull out their satirical rapiers, were disarmed. "That's brilliant," one of them said, twice.

Lear is fit not only mentally but physically. Even in his nineties, he walks in an easy, almost athletic cadence. Exercise is a consistent part of his life, a fact he once demonstrated on *The Dr. Oz Show*. The good physician led Lear over to a yoga mat to demonstrate part of his physical routine. Lear stretched out his ninety-two-year-old body and reached down to his toes. "A three-finger touch!" the host exclaimed. "I used to be able to get my fists down," Lear declared, smiling, "but son of a gun."

In terms of the slowing effects of aging, Lear doesn't have a lot to worry about. And on average, neither do you, if you imitate his lifestyle. The key illustration here is the link between intellectual vitality and physical exercise. One of the most astonishing revelations of recent geroscience is this: greater physical activity means greater intellectual vigor, regardless of age.

Researchers noticed years ago that fit seniors seemed smarter than sedentary seniors, even when wading into the deep end of the statistical pool. Especially powerful were results linking aerobic exercise to changes in executive function. If you survey a large number of studies (called a meta-analysis) looking at aerobics and EF, you see really impressive numbers. Elderly individuals who regularly exercised scored higher, sometimes stratospherically higher, on executive function tests than sedentary controls (effect sizes, which are measures of correlation, were almost seven times greater with exercisers than with couch potatoes). It is quite rare you get such clear numbers in work of this kind.

Yet correlation, as you can hear your logic teacher intone, does not mean causation. To establish that exercise is the cause of the improvements, you have to take a group of elderly individuals with low EF scores, have them exercise for a period of time, and then reassess their EF. If there's improvement, you can tentatively assign the luxurious word "causal" to the experiment.

I am happy to report that such experiments have been done—and the results are consistent and compelling. One study got a 30 percent boost in executive function scores after a skimpy three-month exercise program consisting of an even skimpier "walking regimen." Some studies show much greater improvements. And the boosts appear to be long-lasting. One lab showed that, after people in midlife exercised, their executive function boost was still apparent *twenty-five years later.* Strengthened in the gym of peer review, this idea has muscled its way into our thinking: exercise boosts cognition in the senior brain. No wonder researchers like Harvard's Frank Hu have said: "The single thing that comes close to a magic bullet, in terms of its strong and universal benefits, is exercise."

Naturally, there are the usual ifs-and-buts and what-abouts surrounding such findings. First, not all parts of executive function are susceptible to exercise. The ability to focus, for example, seems impervious to exercise. The effects of exercise on working memory are also mixed. Some studies show a boost if the workout is aerobic; others show no effect at all. The titans of peer review thus state that further research is needed. Not to lose hope, though. Researchers have indeed found something that affects working memory. It appears, however, to be more about the contents of the fork you put in your mouth than about the shoes you put on your feet. We'll have more to say about that in our discussion of diet.

Right now, we need to get into some of the mechanics of why exercise works at all in the brain.

Bulking up your neural tissue

Remember the postapocalyptic Seattle urban metaphor from a few pages back, with brain regions likened to cities, their connectivity likened to highways? Both the structure of the brain's cities and the functions of its neural highways are altered in seniors who exercise. The neural tissue involved in executive function is more active and more bulky, exhibiting a larger overall volume. Scientists readily observe the change in just the area where you really want it: the prefrontal cortex. One particularly sensitive subregion is the dorsolateral PFC, the most connected area in the entire PFC. It is involved in decision making and working memory.

Certain regions in the brain's interior get a cognitive six-pack with exercise, too. Most sensitive is the medial temporal lobe, specifically its crown jewel, the hippocampus. You might recall that the hippocampus is involved in many functions related to clear thinking, including memory and navigation. People who do aerobics bulk up their hippocampal volume by a whopping 2 percent. In contrast, people who just do stretching exercises show a *decrease* of 1.4 percent. People who do nothing, just letting nature take its course, lose 2 percent.

These regions don't just get bigger in aerobicizers, they get denser. In the PFC, it's likely that more connections are occurring within existing neural structures. The hippocampus, however, may be literally growing new neurons, a process called neurogenesis. The protein BDNF, short for brain-derived neurotrophic factor, is thought to be responsible for much of this growth. You want BDNF in your brain. Brain cells take to it like scientists to grant money.

It's not just the cities that grow. Connectivity increases, enabled by the neural cell bodies in gray matter. One study showed an 8 percent increase in global gray matter for seniors who exercised. And the effect was as durable as a tax increase. Nine years later, the exercising group *still* had more gray matter than sedentary controls. Astonishingly, this elevation reduced their risk for dementia twofold.

Given this activity, you might think these newly minted neural structures would need feeding—and need their waste removed—just as the old ones do. And you'd be right. Since both feeding and garbage control involve your blood system, you might predict an increase in blood flow to the new regions. And again, you'd be right. Cerebral blood volume increases dramatically in areas of the brain associated with the exercise-induced growth. The effect is especially pronounced in the hippocampus.

The molecular basis for the improvement in cerebral blood flow has begun to be uncovered, at least in rodents. Exercise stimulates a process called angiogenesis (literally "vessel creation"), and the protein responsible for it is called VEGF—pronounced "vedg-eff," as in vegetables. It's actually short for a tongue twister: vascular endothelial growth factor. It does for blood vessels what BDNF does for neurons. It makes them grow.

But here is what's extraordinary about the data I just described. By exercising, you are not just slowing age-related decline. Your brain actually gets *better* at its job. And you don't have to be an Olympian to reap the benefit. Just take a walk. Or get into a pool. Don't be like "Bootstrap Bill" Turner, from another movie my kids enjoyed watching. It's the third installment of the *Pirates of the Caribbean* series, *At World's End*. In the movie, Bootstrap has been cursed, and we find him nearly lifeless in the bowels of the pirate ship *Flying Dutchman*. He's gradually fusing to the interior walls of the hull, limbs turning to wooden planks, encrusted with sea creatures. For a few moments, he has reason to peel himself off this dangerous hull to talk to his son's fiancée. But it's only transitory. Bootstrap returns to the wall, sedentary once more, the hull resuming its absorption.

Sadly, some people allow the aging process to act like the dangerous walls of the *Flying Dutchman*. They slowly become absorbed into the walls of their years—and into inactivity. If you want to avoid Bootstrap's fate, you must fight inertia. You don't have to do much to get the brain boost. In fact, it may be hard to believe how little you have to do.

A little exercise goes a long way

Research shows you get a cognitive boost with as little as thirty minutes of moderate aerobic activity, essentially walking too fast to talk, two or three times a week. (Some studies recommend thirty minutes five times a week.) The effect is dose dependent—the more you exercise, the better your brain functions—though there is a limit. In one study, seniors walked three hundred city blocks each week; they experienced the welcome increase in gray matter volume. So did seniors who walked only seventy-two blocks each week, however—and by the same amount. Researchers call this a "ceiling effect."

If you add strengthening exercises to your aerobic routine—resistance training for your big muscle groups—you also benefit, regardless of the shape you're in. You have to do strength training two or three times a week as well. Once a week was measured, and that doesn't cut it.

These data act like strong magnets, pulling other recommendations toward them. One is reminiscent of our Bootstrap Bill story. Seniors naturally experience a decrease in mobility as they age. There are many reasons for this slowing down, including reduced energy levels, increased physical pain during movement—even anxiety and depression. Researchers designed a program for people with limited mobility involving aerobic workouts, flexibility exercises, and resistance training. All participants were ambulatory but had limited mobility, as assessed by a test called

. At the end of the program, the exercising group was able to walk about 104 minutes more per week than the controls. And they showed a lot less "major mobility disability." Simply by regularly coaxing Bootstrap Bills to pop out of the immobilizing walls of their lifestyles, they saw positive results.

And that's important. Because we also know that doing even a little bit of exercise goes a long way toward cognitive health—and may even reduce the risk of Alzheimer's. Small incidental experiences of physical exercise are astonishingly effective for seniors, like regularly getting up to cook a meal, walking up small flights of stairs, or going to a movie. Even fidgeting provides health benefits.

One study tracked the physical habits of a group of seniors for four years. The researcher examined limited "range activities," like going for a short walk around the neighborhood, walking out in the yard, or even getting out of their bedrooms. Those who were sedentary were twice as likely to get Alzheimer's as those with "the largest life spaces." Movement even helps people who are wheelchair bound. The bottom line? Shoot for regular exercise—of any kind—even if your body has other ideas. After all, you aren't exercising because you want to move your body. You're exercising because you want to move your brain.

Cheese lovers, beware the bedsheets!

Tyler Vigen's website doesn't seem very provocative at first. It looks like a collection of boring-as-PowerPoint graphs. Each graph consists of two undulating, differently colored lines, looking like Loch Ness Monsters doing some synchronized swimming. In one chart, a line labeled "Divorce rate in Maine" shows a decline from 2000 to 2009. The other line is where things become interesting: it's the "Per capita consumption of margarine in the United States." The two lines are startlingly similar—nearly identical, in fact. The next slide is even funnier. The first line is labeled "Per capita cheese consumption" in the United States. The second is "Number of people who died by becoming tangled in their bedsheets." It lines up perfectly with the cheese, just as Maine divorces stay coupled with the margarine.

What do these slides have to do with this chapter? They are the reason I am reluctant to get into our next subject: nutrition and aging. Like Vigen's slides, a great deal of the published research exploring practical diets for seniors is associative in nature. And as the slides beautifully illustrate, association does not mean causation. Chicken-and-egg problems abound in this work, too. As a result, most of the causal work has been done in lab animals. I have several large problems thinking any of this illustrates something meaningful about *human* aging. That's why I'm reluctant.

But, I hope, not unfair. Research into human nutrition is ridiculously hard—and surprisingly expensive—to do well. Food is complex stuff: even a simple sandwich is composed of hundreds of biomolecules. The metabolic machinery we marshal to extract energy from foodstuffs is many times more complicated than, and as individualized as, a fingerprint. Extracting truth from this pile of variability is like eating soup with a fork. And this complicated field is woefully underfunded.

That doesn't mean research into aging and nutrition is bereft of good—even heroic—work, and we will cover some of the best stuff. To discover where aging fits into the practical world of what's on your plate, we return to the theme of repair breakdown, starting with a very peculiar type of evolutionary gluttony.

Free radicals in a hungry brain

The brain uses a lot of the food it craves for a familiar Darwinian purpose: to project its owner's genes into the next generation. Though it is only 2 percent of your body's weight, your brain consumes 20 percent of the calories you eat. The brain is also quite finicky. It happily extracts energy from sugar molecules, but it turns up its neural nose to fats. If the brain could metabolize fats, you'd literally be able to lose adipose-related weight by simply thinking hard. Unfortunately, the organ is more into sugar than butter, and so taking math tests is never going to be part of anyone's weight-loss program.

As in any normal manufacturing process, the brain generates lots of toxic wastes. Particularly deadly are a few famous molecules humorously (at least for aging hippies) called free radicals. It is important to get rid of free radicals. If allowed to accumulate, they will do considerable damage to the cells and tissues of the body. The damage is termed oxidative stress. Any tissues experiencing oxidative stress in an uncontrolled manner, including neural tissues, start dying. So it's a big deal. Fortunately, your body has an army of molecular defenses designed to neutralize these ever-accumulating toxins. Several prominent battalions in this army are called antioxidants. They get rid of toxic wastes the same way paper towels absorb spilled orange juice. There are many kinds of antioxidants, from proteins you've never heard of like superoxide dismutase to more familiar molecules like vitamin E. As long as antioxidants and other repair-oriented molecular battalions remain on active duty, a balance occurs between towel and spill. The lethal molecular orange juice gets cleaned up, and your body stays healthy.

The sad problem is that as we age, our defenses against oxidative stress literally begin breaking down. Our molecular army goes AWOL for various reasons, divided into our familiar nature and nurture components. The desertion usually occurs in earnest after we've left child-bearing age.

This is really bad news. Those damaging, damnable free radicals accumulate in our tissues, gradually turning our bodies into Superfund sites. It hurts to have such trauma in any part of the body, but it's an especially bad break for the brain, given its extraordinary 20 percent surtax on our energy supplies. The food we eat makes a difference here: keep an eye out in the next few pages for the word "phytochemicals."

Given the connection between the brain and energy from food, it's not surprising that researchers attempting to beat back Father Time have looked at diet. Horace Fletcher in 1913 decreed you could become younger if you simply chewed your food until it was liquid slush. Recommended activity: thirty-two to seventy-five chews per bite. You really can lose weight if you just slow down the process of eating. And since obesity is linked to early death, perhaps old Horace was on to something.

History is replete with the gravestones of people claiming to have discovered the Fountain of Youth. It thus takes a certain bravado for modern researchers to cross swords with our most fantastical mythologies and investigate life extension. Efforts to link healthy aging to food intake can be divided into two groups: the amount of food people consume and the type of food they consume.

Less is more—maybe

It's been observed for centuries that people who eat less seem to live longer—and are oddly happier—than those who gorge themselves. This has striking confirmation in the laboratory, at least if you're a rodent. Severe calorie restriction can lengthen life expectation in certain animals by a whopping 50 percent compared with typically fed controls. The incidence of their many age-related diseases (cardiovascular disorders, numerous types of cancers, neurodegenerative disorders, cancers, diabetes, etc.) goes down—way down—with calorie restriction. The earlier they start, the better the numbers become. Lengthenings have been shown in virtually every animal tested—even fruit flies.

Does this work in *humans*? And if so, should you do it, claiming your 50 percent life-lengthening medal? The real answer is we don't know. There are suggestions that calorie restriction lowers risk factors commonly associated with early death. Consider this research, which involved groups of healthy thirty-seven-year-olds undergoing a 25 percent reduction in their caloric intake—for two long years. Researchers looked at various physiological markers and behavioral traits compared with unrestricted controls.

The results were somewhat predictable—and also extraordinary. The "well duh" result is they lost weight—about 10 percent compared with controls. But they also had declines in blood-based chemicals associated with age-wrecking inflammation (one obnoxious molecule, called c-reactive protein, was 47 percent lower in the dieters). Another unexpected result is that dieters slept better. They had more energy (weird, because they were actually consuming less energy) and were in a better mood (even though they were probably hungry all the time).

These happy findings are associated with longer life, but no one knows if they actually result in longer life. Yet I have a hard time thinking humans would be an experimental exception to nearly every other creature on the planet. It really does appear that what doesn't fill you makes you stronger. If you'd like to try calorie restriction, I suggest you show your physician this page and discuss a plan.

Are you nuts?

Other researchers have looked not at the amount of food eaten but at the type of food eaten. As with restriction research, consistent findings emerge. And it is good news, especially if you've spent your life eating like a native of sunny southern Europe.

I'm obviously referring to the famed Mediterranean diet, so named because it contains ingredients found in Greek, Italian, and Spanish cuisines. The seminal paper was published several years ago in the *New England Journal of Medicine* by, appropriately enough, a Spanish research group. It was called the PREDIMED (Prevención con Dieta Mediterránea) study. The headliner was that people on this diet suffered less often from cardiovascular diseases, including brain-based pathologies like strokes (unsurprisingly, they also lived longer). That gave the researchers a tantalizing idea. Would this diet change other types of brain-health issues besides stroke, say, nonpathological age-related memory loss?

The answer was yes. Though eating southern European food was associated with cardiovascular health, the most interesting result was discovering a big-time arrest of cognitive decline, not associated with cardiovascular issues at all.

These researchers showed many cognitive benefits to the diet, ranging from changes in executive function to changes in working memory. One study randomly assigned three hundred people to three groups: a Med diet supplemented with extra-virgin olive oil, a Med diet supplemented with nuts, and a non-Mediterranean diet. Researchers followed them for four years. Those eating the Med diet plus nuts had composite memory scores a hefty +0.1 above baseline. The Med diet plus olive oil scored +0.04. That may not sound like much, but it's actually huge compared to the controls, which were a depressing −0.17 below baseline. Changes also were noted in frontal cognition scores (essentially executive function) and even global cognition—a kind of gross domestic product test of your ability to think. Both Med diet plus nuts and Med diet plus olive oil scored far above controls here, too. These numbers were obtained from a randomized, intervention-based research design. You don't have to channel your inner statistician to see these are significant findings.

Other studies on the fatty shores of the United States appear to confirm these results. One in particular, called the MIND diet, essentially combined the Mediterranean diet with another program previously shown to lower blood pressure (the DASH diet). Researchers here found not only an arrest in age-related cognitive decline but a reduced risk for dementia. David A. Bennett, director of the Rush Alzheimer's Disease Center in Chicago, wrote about the results of the center's longitudinal studies in *Scientific American*: "[Nutritional epidemiologist] Martha Clare Morris has found that the so-called MIND diet—which is rich in berries, vegetables, whole grains, and nuts—dramatically lowers the risk of developing Alzheimer's."

Bennett's quote provides the answer to a question you may be asking: "What is the secret sauce in all these diets?" Some of the ingredients are familiar, probably sounding like a cross between your mother and your doctor, no cream sauce in sight. Instead, there are tons of fruits and vegetables and legumes. Tons of whole grains, too. And fish, every day. Salt is swapped out for yummy Mediterranean spices.

Some of the recommendations may not be as familiar, however. Nuts are fatty, but they're very much a part of the diet. Oil is the very definition of flab, but in limited amounts—as long as it's from olives—it provides a specific brain boost. The MIND diet is a little different, emphasizing servings of berries and keeping fish consumption to once a week. If you are an American, this is everything you are not already eating. That's why it's called the Mediterranean diet and not, say, the McDonald's diet.

There's still a lot of work to be done trying to tamp down all the hundreds of variables that make scientists like me skeptical. The data still probably boil down to that Ernest Hemingway-esque quote by Michael Pollan: "Eat food. Not too much. Mostly plants."

But these efforts create a nice start. They are the first nutrition studies in years that caused me to sit up and say, "This is worth a look." They form the basis of a series of research projects investigating how the diets work.

Wall posters, oddly enough, turn out to be helpful.

No pain, no gain

Wall posters were trendy when I was an undergraduate. One popular print pictured a bodybuilder lifting weights. As you may know, weight lifting makes big muscles by creating small tears in the muscle fibers, which are then repaired. It is the repair process that provides the bulk. You have to continuously introduce those minor stressors—rips and micro-gashes—in order to look like the poster boy. This is not comfortable—he is actually grimacing in the poster—with the famous caption below: "No pain, no gain!" (A second poster, showing an overweight beer drinker guy eating a cheeseburger, was often stapled over the bodybuilder; its caption was "No pain? No pain!")

This idea of the positive effect of low-level minor irritations—called hormesis—describes something unexpected about diets to fight aging. It actually explains why they work.

Stated biologically, hormesis is the ability to stimulate normal molecular repair mechanisms by constantly stressing the cells carrying them. This includes nerve cells. The stress is always minor, but it's persistent. Annoy them enough times, and the cells start ringing up maintenance, asking the molecular fix-it squads for assistance. These maintenance teams are the very crews that start retiring as we age. Continuously calling them back to service keeps them active, with the happy results that the cells are kept in a better state of repair, the body in better shape, people transiting through old age more comfortably.

Both calorie restriction and plant-based diets exert their anti-aging effects through hormesis, at least in lab animals, and there's increasing evidence of similar mechanisms at work in humans, too. These repair mechanisms fix everything from faulty proteins to leaky cell membranes. They allow extra calcium into nerve cells, strengthening their activities. Certain growth factors become stimulated, such as neuron-loving BDNF (which, as you recall, does many positive growth-related things for the brain's neural constituencies). Dietary restriction stimulates hormesis by convincing the cells that their owner is starving. If calorie reductions are continuously experienced, the repair mechanisms are continuously activated.

Please note that I am not actually recommending you partake of the severe calorie restriction regimens needed to achieve laboratory results. Indeed, researchers show that experiencing such restrictions only five days a month confers the age-related benefits; more than that and you risk negative physiological effects. Not everybody agrees that even that amount is a good idea, however.

Plant-based diets exert their effects because they are filled with so-called phytochemicals, which endlessly tell your brain cells that they are, well, *vegetables*. These phytochemicals somehow persuade those antioxidant armies we discussed earlier to come out of retirement and start taking out the trash, free radicals and all. Combined with the way exercise boosts blood flow, critical for waste removal, you've got a powerful maintenance crew. Phytochemicals also persuade nerves to make more BDNF, jump-starting the process of making new neurons. That the body might think of eating vegetables as an irritating stressor is not lost on this author. But as long as you continue annoying your cells, you also stimulate their life-extending molecules.

We are starting to understand not only what foods you should eat but why they work. And it turns out that the complexity of the foodstuffs we dine on, the aspect of nutrition research that drives me so crazy, may actually be what gives them their anti-aging properties. Consuming individual supplements—like taking vitamin E pills or other antioxidants—doesn't work very well for most people. You mostly just excrete them, which means folks who take a laundry list of supplements just have very expensive urine. The secret appears to be in the synergy between the components themselves, all of which are housed in *actual* fruits and vegetables, many of which are undefined. This makes sense from an evolutionary perspective. Nothing in our eating history gave us the punch of purified supplements, mostly because they don't exist naturally in such powerful concentrations. Indeed, these nutrients were—and are—always tucked inside their plant-based "hosts," and we evolved to eat them the way nature gives them to us. Not the way a pharmacy does. If you want the benefits described in this chapter, you first need to get out there and walk, swim, or just fidget—then go have a plate of phytochemicals.

And make sure it's a small plate.

SUMMARY

MIND your meals and get moving

- Executive function—a suite of cognitive gadgets enabling emotional regulation and cognitive control—tends to fade with age, as the brain's repair mechanisms break down.
- Greater physical activity means greater intellectual vigor (improvements in executive function) regardless of age.
- Though it is only 2 percent of your body's weight, your brain consumes 20 percent of the calories you eat.
- Cutting caloric intake has been shown to reduce chemicals associated with age-wrecking inflammation, improve sleep and mood, and boost energy level—all findings associated with longer life.
- Diets rich in vegetables, nuts, olive oil, berries, fish, and whole grains (such as the Mediterranean diet or the MIND diet) have been shown to improve working memory and lower the risk of Alzheimer's.

your sleep

brain rule

For clear thinking, get enough (not too much) sleep

No one looks back on their life and remembers
the nights they had plenty of sleep.
—Anonymous

I've reached the age where happy hour is a nap.
—Anonymous

"I SLEEP!" SUSANNAH MUSHATT JONES exclaimed, bursting with laughter. She was responding to a reporter who had just asked her a familiar question: the secret to her long life. She also mentioned eating scrambled eggs, grits, and four strips of bacon, a breakfast she'd had every single morning that she could remember.

She had a lot of mornings from which to choose. Jones turned 116 years old in 2015, the last living American born in the nineteenth century and, briefly, the oldest woman alive (she died in 2016). Though she had no children of her own and was married only once—and only for a short while—she had more than a hundred nieces and nephews to spoil. And spoil she did. Ms. Jones put her first niece through college—a terrific investment that grew to include a doctorate—and the niece returned the favor by penning her aunt's biography. Extending her generosity, she started a scholarship fund for African American students. Born to sharecroppers in Alabama, Jones spent most of her life working as a nanny and live-in housekeeper in New York.

Except perhaps for the bacon, Jones lived what most of us would call a healthy lifestyle. She never smoked, didn't drink, and saw her physician several times a year. Remarkably for this day and age, she took only two medications—for blood pressure—and a ream of multivitamins. She was active on her building's tenant patrol team until age 106. The source of her exclamation? She slept ten hours a night, plus naps.

I'll come right out with it: there's more bad news than good news in this chapter. But a little bit of the bad is preventable if we can, like Jones, practice great sleep habits. To understand the effects of sleep on our quality of life in old age, we have to know a little bit about how sleep works, why we sleep, and how sleep changes over time. In this chapter, we'll also talk about the cognitive effects of not getting enough sleep, and finally how to get the best sleep you can. Some scientists believe that sleep, if you're interested in optimizing health in both body and brain, is the single most important experience of the day.

Or should I say, night.

Night owls and early birds

Many people are surprised to discover three things about research on sleep:

1. We don't actually know how much sleep you need per night. Not everyone needs eight hours.
2. Part of a normal sleep cycle involves almost being awakened. Five times a night is *typical*.
3. We're only just beginning to understand why you need to sleep. It's not all about restoring your energy—maybe not even mostly.

These San Andreas–size gaps in our understanding are surprising, given how much experience humans have with sleep. By the time you're eighty-five, you've spent 250,000 hours in slumber-land, about twenty-nine years of your life.

One of the most surprising characteristics of sleep is its hyper-individuality. Many variables affect sleep, making it pretty tough to tell a consistent story.

Country of origin is one example. People in the Netherlands, on average, sleep eight hours and five minutes each night. People in Singapore sleep seven hours and twenty-three minutes. This is how much sleep they *get*. Is that also how much they *need*? Currently, no one knows.

Sleep also varies by chronotype, which is the natural sleep/wake cycle you experience when you smash your alarm clock and wake up when you feel like it. Some people function best if they hit the sheets at 9:30 p.m., then wake up with the morning's overachievers. Some function best if they start sleep at 3:00 a.m. and wake up with the afternoon's rock stars. Other variables include stress, loneliness, and how many sleep-altering substances you regularly consume (coffee lovers?) during the day.

Perhaps the biggest single source of variability is age. Newborns sleep a leisurely sixteen hours a day. Older adults usually get less than six. Even these numbers have to be taken with a giant boulder of salt, however. Some people need five hours of sleep a night. Others can't get by with less than eleven. There was a seventy-year-old British woman who claimed to need only sixty minutes of sleep a night. She was wrong. When sleep scientists examined her over a five-night period, the number was sixty-seven minutes of sleep per night. She showed no obvious behavioral or cognitive disabilities, no sleep-deprivation deficits. While that's unusual, the variability between people is not.

Ease of sleeping is variable, too. More than 44 percent of Italian seniors report serious difficulties sleeping, as do 70 percent of the elderly in France. About 50 percent in the United States and Canada report dozing difficulties. Their problems are divisible into two categories. The first involves getting to sleep, something researchers call sleep onset latency. The second involves staying asleep, something all of us call annoying.

One thing we can say for certain is that sleep quality diminishes with age. To understand how that happens, we first have to understand how sleep works.

The sleep cycle is born of conflict, like two teams competing in the ultimate soccer match: they play each other twenty-four hours a day and don't stop until you're dead.

The sole function of one club—let's dress them in light uniforms—is to keep you awake. This team has lots of talent at its disposal, hormones and brain regions and fluids playing together with a single goal: to keep your eyes open during the day. We collectively call Light Kit the circadian arousal system. Circadian, a word coined in 1959, literally means "about the day."

The other club is composed of a set of biological processes with the opposite goal. Their entire function is to make you sleep. This team—let's dress them in dark uniforms—also involves hormones and brain regions and fluids, but their job is to put you to bed and keep you there for hours. We collectively call Dark Kit the homeostatic sleep drive.

These teams play against each other every minute you're alive, probing, skirmishing, interacting with the enthusiasm of Premier League fans. There's never a tie game, and the play is highly uneven, each team dominating only at certain times of the day. During daylight, the circadian arousal system controls the field. At night, the homeostatic sleep drive rules. Though this give-and-take occurs in twenty-four-hour cycles, it is remarkably independent of sun and sky. The oscillations would take place even if you lived in a dark cave, though then it tends to cycle in about twenty-five-hour chunks, adding an hour for reasons absolutely nobody understands.

Catching the wave

This neurological soccer match—technically called opponent-process theory—can be characterized by brain wave patterns. Brain waves are observable using the hairnet-like EEG device that detects the brain's surface electricity.

The day starts with light uniforms in full control, your brain broadcasting an electrical pattern called beta waves. At night, when the dark uniforms start perking up, these betas are replaced with more relaxed alpha waves, indicative of drowsiness. Eventually you'll be coaxed into a good night's snooze. During the process, your brain descends through three full stages of increasingly deep sleep, the bottom stage occurring about ninety minutes after you've started. This deepest of sleeps, characterized by large, lumbering brain waves termed deltas, is called slow-wave sleep. It's extremely difficult to awaken somebody who is resting on the bottom.

But not impossible. In fact, after an hour and a half, your brain begins to do it for you. The big, slow delta waves give way, and you ascend backward up through the sleep stages, meaning you get "less sleepy." For reasons nobody understands, your eyes signify this arousal by moving back and forth rapidly, a stage appropriately known as REM-1, for rapid eye movement—one. This REM sleep is qualitatively different from deep sleep, logically termed non-REM sleep. At this stage, you can be more easily awakened.

But if all goes normally, you won't be. The dark uniforms will resume their dominance, and you will descend through another three stages of increasingly deep sleep. The large, soothing deltas soon return, allowing you to spend a blissful sixty minutes at the bottom.

That is hardly the end of the arousals, however. The reason it's called REM-1 is because it's only the first of several stages you'll experience that night. You'll typically encounter four more REM events before the night is over, each followed by its own set of deep-sleep dives. Only after the fifth one will the light uniforms pick up their daytime hyperactivity, wresting the field from their opponents and letting you start your morning. This oscillating never pauses for a commercial break. It never stops wanting to wake you in the morning, however much you may resist it, or wanting you to go to sleep at night.

That is, until you start getting older. The teams still want to keep their rhythms, but they find it increasingly hard to do so.

That's the how of sleep. What's the why? The answer seems as obvious as a bad mood. When you don't sleep, you get cranky and irritable, you can't find your car keys or your patience—and most of all, you feel *tired*. Sleep must involve energy restoration, right?

Wrong. Or at least partly wrong. Bioenergetic analysis shows the energy savings during sleep is only about 120 calories, the same as a bowl of soup. And your brain is mostly to blame for this. It's the power hog of the body, taking 20 percent of the energy you consume and required to remain active 24/7 to keep you alive. Saving the energy found in a cup of broth is not impressive. Restoration isn't why we sleep.

Then why do we? From an evolutionary perspective, it's nuts to flatten someone as weak as we are for even ten minutes in the arid plains of eastern Africa, especially in the dark. Yet we regularly stretched out on the savannah, paralyzed for hours, during the same shift normally scheduled for active leopards. Big price to pay for 120 calories.

Only recently have researchers found light at the end of our contradictions. The insights have profound implications for the aging brain. This chapter describes two of the biggest breakthroughs in our understanding of why we sleep.

We sleep to learn (breakthrough one)

The first breakthrough comes mainly from memory research. As you know, your daylight brain is busy recording your various daily activities. Some are forgettable, some are important, and some need time for future processing. Your memory systems are constantly engaged. At least two regions are involved.

The first is the cortex, those layers of world-class intelligence surrounding the brain like wrapping paper. Or a diploma. The second is the hippocampus, that sea-horse-shaped structure we've mentioned often, deeper inside your brain. These two regions form electrical connections with each other during memory formation, communicating like texting teenagers. This activity holds the memory fragments in place until they can be processed at a later date.

What later date? Scientists now know that it means "later that night, during slow-wave sleep." Throughout the deepest slumber, your brain reactivates the memories laid down during the day, the ones marked for later processing. It then repeats their electrical patterns thousands of times, which strengthens connections, consolidating the information they hold. It's called off-line processing. If you can't perform this important reactivation, you can't store anything long term.

Tucked inside these data is a bombshell of a finding. You need to sleep not to rest but to learn. Nighttime is the perfect time for it, when there's little competing information bombarding your brain for attention.

As research continues, it has demonstrated that sleep aids other functions, from digestion to keeping your immune system humming along. Slowly, we're beginning to understand why you need to sleep. It's not because you need to *rest*. It's because you need to *reset*. When resting doesn't function properly, resetting becomes a challenge.

Which, unfortunately, is exactly what happens when you age.

Slow-acting acid

There's a box I keep downstairs, and when I see it I feel despondent. It contains videos of our kids' childhoods.

Why despondent? Not because of the contents—those videos hold some of the most precious memories I have—but because of how the contents are stored. The videotapes are VHS. If I leave the tapes in their current location, I only recently discovered, I might as well store them in slow-acting acid. They'll begin chemically eroding, losing information with the passage of time. This natural degradation doesn't occur immediately and is subject to environmental conditions such as humidity and temperature. But information will get lost—fragmented would be more accurate—if I don't do something. Stored at sixty degrees (assuming reasonable humidity), significant fragmentation will become observable after sixteen years. Increase the temperature to seventy degrees, and loss becomes noticeable in eight. The oldest of our tapes is nineteen years old. See why I feel despondent?

Natural erosion through time is what aging is about, whether you're talking about information stored on magnetic tapes or information stored as cognitive processes. And sleep processes are not immune. In short, they erode, like a VHS tape in your head. Your sleep becomes fragmented.

Specifically, the amount of that memory-inducing, garbage-collecting, completely useful slow-wave sleep (SWS) decreases as you get older. In your twenties, you spend about 20 percent of your nighttime bathing in its healing breakers. By the time you reach seventy, you spend about 9 percent.

To illustrate these changes, consider this comparison between two sleepers during a typical night's rest.

Let's say a kindly grandmother and her twenty-year-old grandson, Noah, go to bed at the same time, around 11:00 p.m. In ten minutes, the grandson has started floating smoothly through the stages of non-REM sleep, surfing the slow waves just before midnight.

Grandma does this, too, but her transit is anything but smooth. She descends the same stages, but upon arrival at the second non-REM stage, she suddenly comes back for air, reawakening around 11:30. Now she has to start the whole process over again. Grandma gets to the same midnight SWS checkpoint, but, unlike Noah, she doesn't stay there long. She comes back up around 12:30 a.m., awakening a second time, and once again has to start the whole process over. She'll ping-pong like this all night, her last visit to the SWS spa occurring around 2:30 a.m., if she gets there at all. Her experience is called sleep fragmentation. Noah, conversely, has cycled smoothly through the entire process, experiencing four to five cycles of non-REM/REM sleep, with four luxurious swims through the slow-wave ocean. He stays asleep the whole night.

What's controlling the sleep experience of both Noah and his grandmother? To explain that, we are going to take a visit to Boulder, Colorado.

The grip of a tiny clock

Buried in the hills of Colorado lies a machine capable of more destructive mayhem than all the world's nuclear weapons *combined*. Here's what would happen if this technology stopped working: civilization would be held hostage. Police, fire department, and emergency medical dispatch communication systems would suddenly go silent. Electrical grids would desynchronize, then overload, creating catastrophic power outages all over the world. Wall Street and attendant global financial sectors would seize up, as if epileptic, and high-speed market transactions would freeze in their digital tracks. Satellite communications would be disrupted, meaning airplanes midflight would no longer know where they were. Neither would you if you were using your cell phone's GPS to get around. That's okay, the phones wouldn't work anyway, except for your previously downloaded copy of *Angry Birds*. Civilization would come to a crippling, grinding, blinded halt.

What doomsday device could possibly hold such ransom over the modern human experience? The answer seems mundane. What's buried in the hills of Colorado is a clock, driven by an engine the size of an atom. The device is the NIST-F2, the world's most accurate atomic clock. It uses the natural vibrations inside a cesium atom to determine exactly what a "second" is, a number required to synchronize most of the world's infrastructure. As long as it functions, civilization flourishes. This powerful chronometer loses one second in three hundred million years.

Buried deep inside your brain is a little patch of neurons—only about twenty thousand cells strong—known as the suprachiasmatic nucleus. The SCN, located several inches behind your eyes, contains the master pacemaker of the body, the cesium clock of human experience. Its natural rhythms are generated—and measurable—through electrical outputs, hormonal secretions, and gene expression patterns. The rhythmic instincts of these cells are so strong, you can excise them from a brain, disperse them into a dish, and they'll still pulse in rhythmic twenty-four-hour cycles. They control what scientists term the human body's circadian system.

And they are the reason it's harder than it used to be for you to get a good night's sleep.

The circadian system works as independently as an entrenched dictator. Yet its scheduling is subject to tweaking—which is one reason we have some control over our sleep. The SCN receives information about the time of day directly from the eyes, along neural trunks called retinal projections. This helps it synchronize its rhythmic output to the turning of Earth. The SCN then uses this information to make you drowsy during the night and aroused during the day. (This function isn't the only factor controlling sleep—core body temperature is also important, for example—and it's not the only thing over which the SCN has sway. The stress hormone cortisol is under tight circadian control. So is digestion. Synchronization occurs because many other biological "sub-clocks" are scattered throughout the body, all communicating with the SCN, like cell phones responding to a cesium clock.)

How does the SCN keep a grip on sleep? This talented nerve knot interacts with many brain regions, including the brain stem, which does most of the heavy lifting in generating sleep cycles. The SCN exerts its rhythmic will via hormones, including its franchise player, melatonin. The hormone is made off-site, a few inches behind the SCN, in a pea-size organ called the pineal gland. During the night, the SCN turns the pineal spigot to "on" and melatonin floods the blood. It'll circulate all night long, not seriously reducing its levels until about 9:00 a.m.

Losing our rhythm

Why does sleep shift from smooth to fragmented as the years roll by? Researchers have uncovered several interesting alterations that occur in elderly brains, all involving circadian rhythms, most involving the SCN.

The aging process does not affect the number of neurons in the SCN. Or its overall size. If you could magically remove the SCN from grandmother and grandson, inspecting only outer structure, you couldn't tell which was which.

That's not true with inner structure. *Most* of the rhythmic systems associated with the SCN are altered with aging. Electrical output changes. The ability to secrete pace-setting hormones diminishes. The expression of rhythm-inducing genes in the SCN declines. All of these have measurable effects on sleep and arousal, specifically targeting levels of melatonin and cortisol. Researchers believe these changes reverberate throughout the body—primarily, of course, in the ability to get a good night's rest. That's why Grandma can't sleep through the night while her grandson Noah glides through it like syrup.

Does this matter for Grandma? Does sleep fragmentation hurt cognition? Researchers used to say yes. The sleep cognition hypothesis postulated that most age-related cognitive deficits could be laid at the feet of sleep loss. But there's a reason it was called "hypothesis." Close investigation revealed that the sleep cognition hypothesis was way too simplistic, bordering on being wrong. Researchers originally thought data that applied to younger populations were transferable to older populations. Two examples will suffice to show the error of this oblique form of ageism.

Memory

Like a song you just can't keep out of your head, the brain replays over and over again at night what occurred in the daytime. We mentioned this overplaying a few pages ago, showing it assists long-term memory consolidation in the human brain. Later investigations showed the boost occurred only for people younger than age sixty. This is thought to occur because of age-related changes in a network of the brain called the corticostriatal network. This network consists of loops that straddle both hemispheres of the brain and are usually involved in mediating feelings of goal-directed behaviors. In older individuals, these loops are just not as active. When researchers assessed off-line processing skills in seniors, using the tests given to younger populations (looking for memory pick-me-ups), they showed none of the benefits less mature crowds enjoyed.

Executive function

Sleep loss is associated with the loss of a number of socially lubricating behaviors, including executive function. That comes from sleep deprivation studies done mostly with willing American undergraduates, and many researchers simply assumed that older populations would show similar deficits. They don't. Sleep deprivation studies in older populations showed no deficits in executive function over baseline—including measures of impulse control, working memory, and attentional focus.

Why wouldn't sleep loss hurt older people? Some researchers believe cognitive deficits due to natural aging don't get any worse because they *can't* get any worse. The damage has already been done. They also can't get any better, for the same depressing reason. This concept is called the floor effect. Cognitive deficits reach a floor below which no travel may be possible.

The situation isn't hopeless, even if flooring turns out to be the wrong explanation. A lesson from the Old Testament points us in the right direction.

An early start

You might recall biblical Joseph, penultimate son of patriarch Jacob, second in command of the sprawling Egyptian empire. He got the position because of the strangest job interview in the world, where he made known his ability to interpret two of Pharaoh's troubling dreams. The first dream involved cows, seven lazy, obese, beautiful bovines, emerging from the Nile to graze on nearby grass. Seven ugly cows soon follow them out of the river (the Bible quotes Pharaoh as saying, "I had never seen such ugly cows!"), lean and scrawny and apparently scrapping for a fight. Straight out of Stephen King, the lean cows turn predatory and carnivorous and attack their fattened colleagues, gobbling them up. The second dream followed the same horror script, but with different characters (involving murderous stalks of wheat). Joseph correctly interpreted these dreams as a warning. Egypt would have seven years of bountiful harvest, followed by seven years of famine. If the people were to survive, they needed to work the fields early and store sustenance for the coming not-so-rainy days.

The job was his.

Here's the lesson for us: saving up for the not-so-rainy days hints at how to treat the effects of age-related sleep fragmentation. If you want to diminish cognitive decline in old age, you must start accruing good sleep habits in middle age.

That's what sleep researcher Michael Scullin thinks. He and a colleague reviewed nearly fifty years' worth of literature on sleep to look for patterns, and they summarized their findings this way: "Maintaining good sleep quality, at least in young adulthood and middle age, promotes better cognitive functioning and serves to protect against age-related cognitive declines."

Storing up good habits now pays dividends when the cognitive famine arrives.

We sleep to sweep (breakthrough two)

Fairly recently, scientists have discovered another, less glamorous function that comes online when you go off-line: garbage disposal.

My research consulting and speaking activities occasionally find me in strange hotels, unable to sleep. From my room, I can watch a given city's night shift: garbage collection trucks loudly rumbling through empty streets, taking waste to landfills; street cleaners rumbling even louder, pushing dirt to the curb. The brain needs garbage collection and street sweeping, too. With all the energy the organ consumes during the day, a lot of toxic waste builds up in its tissues. This needs to be flushed away, just as city garbage must be removed and streets cleaned.

Beautifully, your brain has just such a system. Actually it has *many* drainage systems. Acting like their urban counterparts, many become active at night. One of them is called the glymphatic system. Here's how it works:

Your neurons are bathed in salt-water fluids, similar to the ocean from which they originally sprang. Waste that accumulates in the brain is dumped into these fluids, like irresponsible companies dumping pollutants into nearby streams. Happily, the glymphatic system, composed of cells and molecules and channels, works like a well-funded EPA. It can isolate the junk, remove it from the fluid, and siphon it off to your bloodstream. The toxic waste is removed from the brain, and you pee it out in the morning. This convective system operates during slow-wave sleep, the same stage in which learning occurs.

The same stage that we get less of as we age.

When toxic waste builds up

Even in New York City, a town legendary for labor disputes among its sanitation engineers, the garbage strike of 1911 stands out like rotten meat.

Two organizations, garbage collectors and street cleaners, pressed the city for better working conditions. Officials refused their demands, triggering the strike. It began slowly, with garbage and street debris removed sporadically. As refuse piled high and roads clogged, the city grew increasingly dysfunctional. Officials responded by hiring strikebreakers, who were immediately physically assaulted by the striking workers. Garbage piles piled higher, blocking traffic, creating an incredible stink and a deadly health hazard. To make matters worse, a freak snowfall blanketed the garbage-choked streets right in the middle of the strike. It became in everybody's best interest to get back to work, flush the city clean. That happened a month later, after a great deal of violence, including a few tragic deaths.

Sporadic removal of growing garbage lies at the heart of another conflict, one that takes place in our head during SWS. As you age, your sleep becomes fragmented, and you miss out on this necessary type of sleep. Wear and tear on your SCN—the brain's pacemaker for the sleep/wake cycle—causes the fragmentation. Without slow-wave sleep, researchers believe, the cleaning crew begins to go off-line and waste removal becomes increasingly sporadic.

Just like in the garbage strike of 1911, toxic stuff accumulates. Researchers believe this toxic waste buildup begins damaging brain tissue beyond a certain threshold. The damage includes the sleep apparatus itself, which of course results in more fragmented sleep, less slow-wave slumber, and more damage. Some sleep researchers hypothesize that this damage eventually results in behavioral changes, including cognitive decline and dementia. To summarize, a dysfunctional SCN reduces slow-wave sleep. This leads to only sporadic garbage removal, leading to neural damage.

It's only one idea, and even it suffers from something of a chicken-and-egg issue. Consider that this vicious cycle starts with a dysfunctional SCN and ends with dementia. It could be that the molecular garbage pile begins accumulating for reasons other than sleep loss (a genetic origin is a distinct possibility). And then it isn't until toxic waste reaches that certain toxic threshold that the SCN is rendered dysfunctional. This triggers the rest of the steps. At this stage of our understanding, researchers are not sure if the SCN starts the ball rolling initially or jumps in later in the game.

Where does this hypothesis originate? Researchers have known for years that a chronic lack of sleep is a risk factor for many neurodegenerative diseases—including Parkinson's, Huntington's, and Alzheimer's. Consistent with this epidemiological observation, it was noticed years ago that jet-lagged flight attendants (especially those logging long-haul international flights) have an unusual amount of hippocampal atrophy, a telltale sign of Alzheimer's. Eventually researchers showed that circadian disruptions (in any profession) promote system-wide inflammation and uncleared toxic waste.

Those convinced that the amyloid hypothesis explains Alzheimer's use this hypothesis to bolster their claims, and for this reason: it is now abundantly clear that the inability to flush out the toxic amyloid fragment Aβ is what really causes the damage in Alzheimer's. With increasing sleep loss, the fragment appears to hang around longer than it should. That's why lack of sleep is a risk factor for Alzheimer's. Add in the fact that the glymphatic system dramatically slows whenever you awaken and a consistent story emerges: Aβ has no way of being consistently removed. Dementia is the hypothesized result.

This alone provides a powerful reason to get a good night's sleep, and at any age. But it's hardly the only one. The length of your life and your mental health are other solid arguments for why sleep is important, and it is to these issues that we turn next.

The bedtime story that's just right

For surprisingly professional reasons, many scientists *love* the Goldilocks story. It gives us a way to illustrate an interesting tendency common to many of the biological processes we study. Behavioral processes, too.

My favorite Goldilocks variant comes from the old *Rocky and Bullwinkle Show*. This version told the story of blond Tussenelda Woofenpickle (whose nickname was "Goldilocks," intoned the narrator, because of her golden curls). Goldilocks gets lost in the woods and finds the Bear Family cabin populated by Mama, Papa, and Little Oswald. Only Oswald's stuff, from porridge to rocking chair to bed firmness, is "just right." The parental units' stuff either overshoots or undershoots Goldilock's delicate sensibilities. Edward Everett Horton's stentorian voice-overs, accessorized in a pseudo-British/ Brooklyn accent, give the episode a satirical, false gravitas. After all these years, it's still fun to watch.

Still instructive, too. I discuss here the optimal number of hours you need to sleep to get the best shot at the highest quality of life. And the highest probability of living the longest while you're busy enjoying this quality. As you'll see, the data follow an inverted U shape: two extremes that don't suffice and a sweet spot in the middle that's just right.

Studies show that sleep disruptions aren't just inconvenient. They're deadly. Not getting a specific amount of sleep affects how long you live. From studying thousands of people (21,000 Finnish *twins*, actually), we can even say what that amount is.

Here's the bottom line: you need to get between six and eight hours of sleep every night, no more and no less. If you get less than six hours, mortality risk rises 21 percent in women, 26 percent in men. If you get more than eight hours, mortality risk rises 17 percent in women, 24 percent for men. You have to have the "just right" amount of sleep to optimize both quality and quantity of life. Sound familiar? This is where our Goldilocks discussion becomes relevant.

The mortality risk is referring to any cause of death. But not surprisingly, the usual suspects are the ones associated with old age: stroke, heart disease, blood pressure issues, type 2 diabetes, obesity. What is surprising is that these numbers are actually lower than they are for younger people. Young men, for example, see a 129 percent greater mortality risk with sleep loss. How does *that* work, and why is there a difference between the generations? Currently, we have no idea.

And you'll have to take these figures with a grain of statistical salt. Yes, the data are solid. However, as you may remember from high school math, statistics don't apply to individuals. Sleep duration requirements still vary from one person to the next.

I mentioned near the beginning of this chapter that many seniors, across a range of countries, report difficulty sleeping. That's important for several reasons. One sleepless night can leave you grumpy, but several in a row can leave you cognitively impaired. Everything is affected, from memory functions to problem-solving abilities.

Worse, there's a deeply troubling association between consistent sleep loss and mental health. Seniors taking more than thirty minutes to fall asleep have an increased risk for anxiety disorders, and for a reason probably familiar to you. They begin reviewing all their troubles at bedtime, experiencing an endless film loop of worry, replaying the same concerns over and over again. Though this caustic habit of rumination can afflict any age, old people have uniquely serious reasons to worry. They may be feeling a lack of control over mind and body, especially if medical issues are involved, and uncertainty about finances and relationships. Soon thirty minutes have gone by, and all they've got to show for it is a sweaty bedsheet.

Depressive disorders are also linked to increased sleep fragmentation. Seniors who suffer from depression generally fall asleep quickly. They just don't stay there. Depressed elderly get the worst sleep of all.

Why does this troubled business partnership exist between sleep and mental illness? We have no idea. Though we know sleep and affective disorders are tightly linked, we don't yet know the direction of the linkage. Fortunately, that hasn't stopped researchers from trying to figure out ways to help us all sleep better. One who rolled up his practical research sleeves was the late sleep scientist Peter Hauri.

How to get better sleep

Dr. Hauri had a German accent as thick as bratwurst. Swiss by birth, he was blessed with a Matterhorn-size laugh and a mind like a Rolex. He got into sleep research after immigrating to the United States, where he quickly rose to prominence. For years he headed the Mayo Sleep Disorders Center in Rochester, Minnesota.

Some of his research ideas made headlines. He suggested people get rid of their alarm clocks. He encouraged insomniacs never to *try* to sleep, declaring it just makes them more aroused. And he recommended people keep track of their sleep habits the way some people keep track of their diets. His ideas were eventually codified into the book *No More Sleepless Nights*. It was for years the go-to book for treating insomnia.

Some of Hauri's insights, along with more recent findings, are listed below. You need to adapt them to your own situation, however. Hauri would be the first to tell you that everyone's sleep habits are unique—"like snowflakes," he once said, undoubtedly with a twinkle in his eye.

1. Pay attention to the afternoon.

Getting a good night's sleep starts with paying attention to what you're doing four to six hours before you go to bed. No caffeine six hours prior. No nicotine. No alcohol, either. Alcohol, legendary for inducing drowsiness, is actually a biphasic molecule possessing both sedating and stimulating properties. Drowsiness occurs initially; the stimulating effect much later. When you drink, you spend less time in REM and SWS, especially in the terminating night hours. Exercise has a profoundly positive effect on your ability to sleep, but you want to do it earlier in the day. It's become evident in recent years that ensuring a good night's sleep occurs way before you hit the sheets.

2. Create a sleep "terrarium."

Designate a place in your house where the only activity is sleep. That's going to be the bedroom for most people. Don't eat there, don't work there, don't have a TV there. Just sleep. (You might do one or two other activities there, but see the above admonition concerning exercise.)

3. Watch the temperature.

People fall asleep ideally around 65 degrees. Make sure the room you just designated as Sleep Central is cool. Install a fan if needed, which is a good idea for another reason. In addition to temperature regulation, it provides steady white noise. That helps many people go to sleep.

4. Create a stable sleep routine.

Go to that cool single-use bedroom of yours at the same time every night. Wake up at the same time every day. No exceptions. If you're unable to fall asleep in time to get six or seven hours of sleep at first, continue to wake up at the same time, to reset your routine.

5. Pay attention to your body's cues.

If at all possible, don't bed down until you're tired. And if you wake up during the night, don't turn the experience into an Olympic toss-and-turn fest. If you can't fall asleep after thirty minutes, don't stay in bed. Get up and read a dead-tree (non-electronic) book. Especially one that's boring.

6. Pay attention to light exposure.

Expose yourself to bright light during the day, dim light during the evening. This mimics what our brains were used to experiencing during our sojourn under the vast African skies.

7. Stay away from blue light.

That means laptops, TVs, mobile devices, or anything that radiates at 470 nanometers (the frequency of blue light). That wavelength has been shown to trick the brain into thinking it's daylight. Arousal follows, and for a logical evolutionary reason. Blue is the color of sky, which the brain historically encountered only in the daytime.

8. Visit lots of friends during the day.

Depression is associated with sleep fragmentation, and social interactions are powerful antidepressants. Social interactions also exert a surprisingly powerful cognitive load, giving the brain a real workout, preparing it for surfing the slow waves later that night.

9. *Keep a sleep diary.*

This is especially important if you have serious problems sleeping and are considering professional help. A simple version involves documenting when you wake up, when you go to sleep, and your frequency of nighttime awakenings. You can find more sophisticated templates online. (Hauri's book *No More Sleepless Nights* also has templates you can use.)

Most of these suggestions are settled law, many coming from Hauri's work at Mayo. However, everybody's situation is different. We've covered most of the basics, but we've neglected certain environmental issues, such as debilitating pain, and all nature issues, such as genetics. But there's one specific issue I want to address: insomnia.

A few years before Hauri's death, a protocol was tested with the aim of helping seniors with troubled sleep. University of Pittsburgh researchers developed the protocol, called the brief behavioral treatment for insomnia.

The intervention was simple. Researchers first obtained a "sleep baseline" from each senior. Behavioral and physiological measures included actigraphy (involving a wearable sensor that measures motor activity) and polysomnography (involving recording brain waves, cardiovascular activities, and more). Then the seniors took a mini-class explaining how sleep works, including opponent-process theory. They were introduced to their research task:

1. Subjects were to reduce the amount of time they spent in bed (six-hour minimum).

2. Subjects were to observe a strict adherence to a daily schedule, arising from bed at the same time—even if their previous sleep was of low quality.

3. Subjects were not to go to bed until sleepy, regardless of the time.

4. Subjects were not to stay in bed for long if they had not fallen asleep.

Teaching these ideas took about an hour, with a thirty-minute "refresher course" two weeks later, and the instructors called the seniors a couple of times to check in on compliance. At the fourth week, subjects came back into the lab to retake the tests.

The idea was to get their sleep schedule to run like a watch, breaking the hold insomnia had on these elderly folks.

It seems like a small effort, but don't let that fool you. Fifty-five percent of the group who underwent the treatment showed no insomnia by the time they were finished. That's complete remission, folks, from a formerly *very* sleep-troubled population. And six months later, many continued to see the positive results: 64 percent maintained a dramatic improvement in their sleep experiences—and 40 percent were still in remission from insomnia. What's interesting about these data is what's missing. There was no psychiatric counseling. There was no sleep-inducing medication. (That's a good call. In older populations, the side effects of commonly prescribed sleep sedatives tend to be overwhelming. And sleep gets only marginally better.)

This protocol is a great example of a theme we've visited many times: the power of lifestyle changes in combating the negative effects of aging. "Lifestyle changes" means changes in lifelong habits. So the practical suggestions here, if adhered to closely, have long-term, life-affirming consequences.

So far in this book we've talked about ways to improve the quality, and maybe even length, of your life. There's one question you've surely thought about if death is only a decade or two away. Can you arrest the process of aging? Can you give it a speeding ticket, convince it to slow down, maybe even stop it altogether? We are going to discuss attempts to increase longevity and, in so doing, separate one last time science from science fiction.

SUMMARY

For clear thinking, get enough (not too much) sleep

- Scientists don't actually know how much sleep you need per night. Nor do we fully understand why you need to sleep.
- The sleep cycle is born of a constant tension between hormones and brain regions vying to keep you awake, and hormones and brain regions trying to make you go to sleep. This is called opponent-process theory.
- Sleep, we are finding, doesn't have as much to do with energy restoration as it does with processing memories and flushing out toxins in the brain.
- As you grow older, your sleep cycle becomes more fragmented, particularly the part of the cycle during which toxins are flushed out of the brain.
- Accruing good sleep habits by middle age (a stable sleep routine; no caffeine, alcohol, or nicotine six hours prior to going to sleep) is the best way to avoid sleep-related cognitive decline in old age.

FUTURE BRAIN

your longevity

brain rule

You can't live forever, at least not yet

Millions long for immortality who don't know
what to do with themselves on a rainy Sunday afternoon.
—English novelist Susan Ertz

I don't want to achieve immortality through my work;
I want to achieve it through not dying.
—Woody Allen

YOU KNOW THOSE SPRY next-door neighbors, the eighty-year-olds who live in stand-alone houses and cut their own lawns and are so bright and facile you can practically see their minds from space? They're sometimes called "Super Agers," groups of elderly who neither think nor act their age. When you test their memories, they score more like fifty-year-olds than eighty-year-olds. And they tend to live much longer than average folks, too.

What can Super Agers teach us about why people live as long as they do? Even more irresistible, how long could "long" be? Researchers and crackpots alike have wondered for centuries—and still do to this day.

For example, you have people cryogenically preserving their heads, waiting for a time in the hazy future when scientific knowledge will be advanced enough that their heads can be (a) thawed without damage and (b) restored to conscious existence somehow. And you have at least one guy running for president, in 2016, on an "immortality platform." He rode around the country in an RV decked out like a coffin, with "Immortality Bus" painted on the side. The candidate explained, "I'm a firm believer that the next great civil rights debate will be on transhumanism: Should we use science and technology to overcome death and become a far stronger species?" As a scientist, I'm flattered by the faith they put in my profession, misplaced though it may be.

Great strides have been made in our understanding of the finicky biological springs and gears that allow Greenland sharks to last five centuries but allow us to last only one. Serious scientists, tinkering with the flywheels behind aging and longevity in laboratory animals, have had remarkable success extending the lives of these creatures. There are also nonscientists using sketchy research to make stupid claims about living forever. They aren't tinkering with anything except the truth. In this chapter, we're going to look at the great strides.

First, I want to clear one thing up. Aging is not a disease any more than puberty is. It's a natural process, one that usually leads to a whopping misunderstanding. People don't *die* of old age. People die of discrete biological processes that break down because they've spent too much time on the planet (for most people, the weak link is their cardiovascular system). It's thus not surprising that scientists do not recognize aging as a pathology. That's why you usually don't find researchers trying to find a "cure" for it. Rather than trying to discover why things go wrong, they're trying to discover why things go right.

Different question. Much more interesting answer.

For some reason, many of the best studies exploring this question are British. These expensive longitudinal studies follow people from birth to the present day, tracking everything from how people's physiologies fare to how well their minds hold up. One study, the National Survey of Health and Development, was started in 1946 to follow the life histories of over five thousand people. Like a British Energizer Bunny, it's still going. Another is the National Child Development Study, tracking the life histories of seventeen thousand Brits born in 1958. One of the biggest is the Millennium Cohort Study.

It's captured nineteen thousand in its research net, tracking people born between the years 2000 and 2002. That makes it one of the "babies" in this odd British family.

Clear patterns have emerged from this work. One consistent finding concerns those spry next-door neighbors.

Researchers have gazed into the brains of these aging overachievers using noninvasive imaging, and what they find is both astonishing and consistent. Their tissues don't look anything like the brains of your typical octogenarians. Their cortices are still thick and lively, particularly in an area known as the anterior cingulate. This region is associated with cognitive control, emotional regulation, and conscious experience. These changes percolate up to the surface as measurable behaviors. Scientists often call these nimble neighbors the "Wellderlies."

Their cognitive performance appears to be genetic. For example, one Scottish study tested childhood IQs at age eleven in 1932, then again at age seventy-seven. They found senior cognitive performance was predicted by only one factor: how intelligent you were in 1932. To quote a research geneticist, "The participant's score at age eleven can predict about 50 percent of the variance in their IQs at age seventy-seven." That means performance measured at puberty can predict with astonishing accuracy performance six decades later. No other factor comes close: not external activities, not level of education, not physical activity, *nothing.*

Could longevity be carved into our DNA, too? Other researchers say yes, albeit shyly. Several studies have found that (a) longevity is determined by the contribution of many genes ("polygenic") and (b) there may be a hierarchy, with some genes playing more of a leading role than others. All told, anywhere between 25 percent and 33 percent of the variance in life expectancy can be explained by how well you chose your parents. Wellderlies have an especially strong genetic component. If you have lots of centenarian relatives, you're more likely to be one, too.

What does this mean for the rest of us? The existence of Wellderlies and the stability of certain traits regardless of age give researchers a rational basis for asking if there really is a Fountain of Youth. If you can find the secrets to why some people live as long as they do, perhaps you can figure out how to extend life in others. This remarkable feat has been accomplished in lab animals—and it's actually not all that hard to do.

The Holy Grail of Indy genes

I'm not sure if Monty Python knows this, but the comedy group has a gene named after it. The honor was bestowed for a scene in *Monty Python and the Holy Grail*, in which a victim of the plague, slung over a shoulder to be carted away for burial, says, "I'm not dead yet!" An argument then ensues—with him—over whether he's really dead. The gene was first found in a fruit fly, and it actually extends the life of the insect.

The isolation of this gene, by Stephen Helfand, was made possible in part by the canonical 1970s work of Michael Rose, and it all had to do with sex. Rose took seriously the fact that natural selection loses interest in us after our mating days are finished. Rose asked this question: What if you took a bunch of fruit flies and didn't let them mate until they were at an advanced age? (In the case of fruit flies, that's about fifty days, so answers can be gained quickly.) Only those fruit flies hardy enough to survive would be able to project their genes to the next generation. Those animals that couldn't mate couldn't contribute eggs. If you did such "age selection" over a large number of generations, could you generate fully reproductive older animals living longer lives? Rose had to wait only twelve generations to get his answer. His selected insects *were* living longer. He eventually created populations of animals he calls Methuselah flies, which routinely last 120 days.

These data acted like a match to a research fuse. Progress exploded, and the work into life extension became more detailed and exacting—

which is where Monty Python fits in. Scientists eventually found the gene in the insects that, when mutated, conferred long life without having to wait a dozen generations. It was christened the Indy gene, the acronym for "I'm not dead yet." An ingenious name for a longevity gene.

Fruit flies weren't the only creatures researchers successfully turned into biblical characters. Nowadays you can get similar results in many laboratory specimens, from yeast to mice. Mice are the most important because they aren't just vertebrates; they're mammals. Like us.

The work in mice started with dinner. Or actually, a lack of dinner. Scientists observed that calorie-restricted rodents lived longer than regularly fed ones, something we discussed in the exercise chapter. Researchers hypothesized that genes involved in growth and metabolism might also be involved in longevity. Typically, a mouse lives about two years. Researchers wondered if they could bump up that number by interfering with specific genes.

With "knockout" genetic engineering technology, they were able to show it. Researchers create a lab mouse that is typical in every way except that one gene is dysfunctional—literally "knocked out." Their target was a growth hormone receptor in a dwarf mouse called, unceremoniously, GHR-KO 11C. The animal lived past its second birthday and kept going. By the time the lab celebrated the animal's fourth birthday, researchers knew they had something special. Yet they didn't know how special it was going to be. GHR-KO 11C lived almost twelve more months, dying just short of its fifth birthday. If that animal had been human, it would have lived almost 180 years.

Researchers know how to extend the life of many familiar lab denizens now. One creature, a roundworm with the tongue-twisting name *Caenorhabditis elegans*, has met with especially spectacular success. Mutating a gene called age-1 can extend its life to more than 270 days. That's amazing, considering it usually only lives about 21 days. If that animal had been human, it would have lived to almost eight hundred years.

As far-out as that might sound, it's nothing compared with what cancer cells can do.

The cells of Henrietta Lacks

If someone had told me that cancer cells I used to work with as a postdoctoral fellow would eventually be honored by Oprah Winfrey, rock the leadership of the National Institutes of Health, and trigger legal action involving one of the world's top research publications, I wouldn't have believed them. And if *I* told you these cells actually came from a woman who had died years before my birth—yet kept dividing at such a robust rate we had to isolate them from other cells in the lab to guard against contamination—you might not believe me, either. This, however, is exactly what has happened. The cells are called HeLa cells, and they are some of the most famous human tissues in the world.

HeLa cells have an origin as humble as Winfrey's. They belonged to Henrietta Lacks, a tobacco farmer from Virginia. Lacks moved to Maryland in her later years, where she was diagnosed with the cervical cancer that would claim her life. Doctors removed samples of her cervical tumor—without her permission—during the course of treatment and gave them to research scientists. That lack of permission eventually led to the notoriety I mentioned. The investigators put her cells into glass dishes with a kind of nutrient broth—tissue culture, it's called—in an attempt to understand how cancer works.

Henrietta Lacks died in 1951. But her cells did not. Unlike other tissue culture cell lines of the time, her cells shockingly continued to grow and divide. They still do, which is why I, as a young scientist, could use them decades later. They're quite hardy. Scientists froze HeLas, thawed them, divided them, mailed them to other scientists, and, with proper care, grew them indefinitely. It sounds like a fantasy, but scientists say Lacks's cells have been immortalized. We now know that many human cell types, as long as you're ghoulishly willing to make them cancerous, can be immortalized.

Yep. *Immortalized.* And you can bet researchers have been stumbling all over themselves trying to find out why.

Counting tips

The solution comes partially from a scientist blessed with industrial-strength brilliance. Leonard Hayflick, a legendary researcher on aging, was the first to show that healthy cells die in culture because they have a molecular accountant keeping track of how many times they've divided. Once they've crossed a certain threshold of divisions, the accountant tells the cells to stop dividing, leading to senescence and death. The threshold beyond which a cell no longer is allowed to split is called the Hayflick limit.

This accountant is as keen as an IRS auditor. Even if you let cells grow for a period of time, freeze them, and then thaw them so they can resume dividing, the cells don't reset to zero and enjoy a fresh set of replication permissions. They continue counting from the point where they left off. Hayflick suggested this accountant be called a "replicometer."

His work has given birth to many research questions. Are cells immortalized because they did a whack job on the replicometer? If we could isolate the replicometer, might we have an important key to finding the molecular basis of longevity?

Such a replicometer has indeed been uncovered. And its elucidation garnered a Nobel Prize for the scientist who found it. That wasn't Hayflick but a neighbor of his, a colleague just across San Francisco Bay. How does it work? Bear with me as we review some biological concepts you might not have studied since high school.

As discussed, a typical cell's nucleus houses the encyclopedia of *you*, written in the dialect of DNA. That DNA is divided into "volumes," a total of forty-six, each volume called a chromosome. At a certain stage in a cell's life, those forty-six chromosomes look like little *x*'s. The nucleus then looks like a bowl of alphabet soup (if the only letters were *x*'s).

The tips of chromosomes turn out to be extremely important to our cellular survival story. They're made of special structures composed of DNA and glops of protein. The whole thing is called a telomere. The DNA at the telomere is made of hopelessly repeated segments; the protein serves mostly to get in the way of a very important function, as we'll discuss in a minute.

Like all living things, cells like to reproduce, though most do it in an unarousing, asexual fashion. This process is called mitosis. Mitosis begins with a cell copying its DNA, which means copying its chromosomes. Tiny little Xerox machines do the job, zipping along the length of a chromosome, faithfully duplicating what they see until they reach the end. Finished, the cell splits down the middle, creating "daughter cells." A copy of each duplicated chromosome gets into each daughter.

There's only one annoying problem with the copying part. When the Xerox gets to the chromosome's tip, it bumps into that gloppy telomere. The machine gets stuck and can't reproduce that tiny last bit of DNA. What does it do? The machine gives up and falls off. The last bit of DNA tip is *not* replicated. This surrender is as constant as paper jams. It occurs on all chromosomes and happens every time a cell reproduces itself. Since some cells reproduce every seventy-two hours, the tips get shorter and shorter by the week. Researchers now know that this serial amputation serves as a kind of doomsday clock: when enough of the tips have been lopped off, the cell gives up and dies.

This countdown forms the basis of the Hayflick limit. It is part of the replicometer. And it may explain why we hang around planet Earth for only so many years.

Help is on the way, sort of

The cell knows about this ticking clock like an inmate on death row. Given the stakes, you might guess cells would come up with some form of checks and balances to ward off such deadly tip erosion. You would be right. Many cells possess an enzyme (the "verb" name for a protein) called telomerase. Its sole job is to find the molecular stumps of chromosomes and add back the tip, filling them in with "prosthetic" telomere sequences. But telomerase, like the federal government, doesn't work very well. So most cells retain their death clocks. That's actually a good thing. If telomerase were allowed to add back tips whenever it saw a stump, there'd be no "time's up" signal. Cells would replicate in an unrestricted fashion and, as long as they were given enough food, never die. They'd be *immortal*. We have a name for cells that replicate uncontrollably. We call it cancer. You can see why I could work with Lacks's cells more than half a century after she died. Cancer made cellular death optional.

As I said, you might want to be thankful most cells don't allow telomerase unrestricted free agency (it's not even *available* in some cells). The consequence, however, is cellular mortality, tissue mortality, and finally your mortality. This gives rise to an odd fact. In the twisted logic of biochemical survival, death is nature's way of keeping you from getting cancer.

There was a time when we thought telomerase might be the key to longevity. When its function was first discovered, speculation abounded that if we fiddled with it long enough, we might get more life. Attempts to confirm this idea failed. What we mostly got was more cancer.

It's important to understand the telomere and telomerase. Elizabeth Blackburn et al. won a Nobel Prize for finding out what they do. Longevity and telomerase probably do have some relationship not yet understood. But when it comes to the complexity of longevity, we are nowhere near using genetic techniques to make some of us live to our five hundredth birthday. We're still working on how to get most of us past our one hundredth.

The rise and fall of longevity genes

The historian Edward Gibbon has lessons for us on complexity. Sickly as a child, sad victim of a parent-vetoed love affair as an adult, he turned his back on his painful present and focused his considerable intellect on the past. The ancient past. Gibbon became an expert on Roman history, publishing several legendary volumes around the time of the American Revolution. His most famous work was *The History of the Decline and Fall of the Roman Empire*. Gibbon's central thesis was that Rome didn't fall all at once, floored by some massive imperial heart attack. Rather, it hemorrhaged to death from the cumulative effects of thousands of small sociopolitical pinpricks. These punctures ranged from collective self-centeredness (the citizens lost something he called "civic virtue") to military weakening (the defense duties were outsourced to uncommitted mercenaries) to Christianity (the hope for a better life led to a disinterest in the present one). These cultural paper cuts, in his view, slowly bled life from one of the largest empires of its time. It then died of exhaustion.

The contributions ultimately responsible for aging and longevity are just like Gibbon's central thesis. Our decline and fall come from the cumulative action of many randomly deteriorating processes. They're counterbalanced, futilely, by the cumulative contributions of longevity genes, telomerase possibly among them.

I want to mention a few other genes (of many) that make vital contributions to the longevity narrative: sirtuins, insulin-like growth factor 1 (IGF-1), and the mTOR pathway.

Sirtuins

This aristocratic-sounding family of proteins has members that, if coaxed to overproduce, will lengthen life in the usual suspects: yeast, roundworms, fruit flies, and mice. Mice that overproduce sirtuins, for example, are better at resisting infectious disease, have more physical endurance, and show improved overall organ function.

There's good news even if you're not a mouse. You don't have to rely on genetic engineering to compel sirtuins to overproduce. Ingesting exotic-sounding biochemicals like chalcones and flavones and anthocyanins and reservatrols also do the trick. The first three of these molecules are found in fruits and vegetables, the last in wine. Scientists speculate that diets like the Mediterranean and MIND—washed down by grape-based alcohol—work because they're as full of vegetables as they are of data.

IGF-1

This gene, whose full name is insulin-like growth factor 1, confers life-extending properties on creatures by *under*producing itself. Unlike with sirtuin, the less you have of IGF-1, the longer you'll live. Notice that I said the word "you." This is because this finding has been widely demonstrated in human populations. The title of the first paper discovering this says it all: "Low insulin-like growth factor-1 level predicts survival in humans with exceptional longevity."

Further research showed these life-extension effects were as selective as Title IX. Underproduction of IGF-1 can predict long life in women but not in men, except under one unfortunate circumstance: if males already have a history of cancer. Only then does IGF-1 reduction become an equal opportunity gift. Given the name "growth factor," overproduction leading to cancer is not all that surprising.

mTOR pathway

The last one is interesting both for its structure—note that it's called a *pathway*—and its cellular job description. This pathway is actually a group of molecules containing proteins that function as part vitamin, part psychiatrist. mTOR promotes growth—there's the vitamin—but it's also involved in responding to stress once the cell encounters something stressful—there's the psychiatrist. Reducing this pathway's ability to signal, thus inhibiting its twin job descriptions, somehow increases longevity in lab creatures. Like sirtuins, it is a friend with health benefits: it can power up immune function and halt age-related cardiac decline.

Recently, researchers have discovered a way to decrease this pathway's activity, no genetic engineering required. All one need do is take a pill. You did not read that wrong. There's a pill you can give lab animals that extends their lives. The active ingredient is rapamycin, an immunosuppressing antibiotic that also moonlights as an anti-cancer drug (there's that pesky cancer/longevity connection again). It specifically interacts with the mTOR pathway, extending life span by about 30 percent in female mice.

A pill for aging?

Rapamycin is hardly the only pill under investigation—nor is the twenty-first century the only era to tinker with ingesting chemicals to find the Fountain of Youth. Writing for *Time* magazine, journalist Merrill Fabry created a delightful time line of this historical contest. One ancient Sanskrit text declared the way to extend life was to eat a yummy concoction of butter, honey, gold, and some kind of root powder. It was to be taken right after a morning bath. No less a luminary than Sir Francis Bacon also suggested baths to extend life—administered along with a healthy dose of opium. Physician Charles Gilbert-Davis, writing in 1921, outlined incredible results from intravenously giving patients small doses of radium. This is the cancer-causing element that killed its discoverer, Marie Curie, who died of aplastic anemia because she kept the chemical in her pockets. So much for life extension.

Some ancients declared it wasn't what you put in your mouth that provided the benefit, but how it got there. An ancient Chinese alchemist advised Han dynasty emperors to use cutlery made only from gold when they ate. The gold had to be extracted from cinnabar, unfortunately—a nasty requirement, for it also contained a toxic compound of mercury.

While these suggestions might sound silly these days, we dismiss these forebears at our peril. Some of their ideas would prove valuable later. Many twenty-first-century researchers are still in the pharmacological race for long life. I outline below several well-known drugs currently under investigation by reputable labs or marketed by reputable companies. They are all seeking to win the longevity competition. First prize, if they ever succeed, would be worth trillions.

Metformin

This drug demonstrates the power of dumb luck in science, for it was originally approved by the FDA to treat diabetes. Several years back, a group of researchers were doing epidemiological studies on metformin's potential long-term side effects, and they noticed an odd thing. People who took it lived longer than nondiabetic controls. They also had fewer strokes and heart attacks, perhaps related to the longevity finding. Their rates of cognitive decline slowed considerably, too. Further investigations showed that metformin worked on the mitochondria of cells, small structures that act like batteries in a smartphone, supplying energy. Metformin's potential life-extending properties in humans are currently under intense investigation.

Montelukast

This one isn't so much a whole-body longevity drug as a whole-brain longevity drug. It profoundly affects age-related cognitive decline in rats. With animals suffering dementia (yep, those exist), montelukast has been shown to orchestrate a near-complete restoration of cognitive function. It's thus an anti-aging strategy particularly suited for the brain. This hasn't escaped the notice of the legions of researchers interested in arresting neurodegeneration, of course. Montelukast exerts its effects by targeting leukotrienes, biochemicals normally involved in mediating inflammation in the human lung. What this has to do with cognitive extension is a complete mystery.

Basis

One pharmaceutical that's gotten lots of attention from the press is marketed by a company called Elysium Health. The visibility comes mostly because the company has no fewer than six Nobel laureates sitting on its advisory committee. The product is a little blue pill called Basis, made from, among other ingredients, extracts of blueberries.

The active ingredient in Basis comes from a naturally occurring biochemical called NAD (short for the nicotinamide adenine dinucleotide), known to extend life in mice. Remember that family of life-extending genes called sirtuins? NAD is the molecule upon which the proteins encoded by the sirtuin genes act, permitting certain metabolic processes to function effectively. Unfortunately, NAD levels drop with age. If you could boost those levels, could you extend your life? No one currently knows. It's being marketed as a supplement, which avoids FDA scrutiny, which has led more than a few scientists to wrinkle their noses at any anti-aging claims. To be fair, the people who run Elysium do, too. They simply say it's aimed at "cellular health." Aging, after all, isn't considered a disease.

Sigh. As with all these efforts involved in creating pills for aging, much work needs to be done.

Blood brothers

Like hereditary titles granted in reverse, many ancient cultures believed youthful vigor could be physically transferred to old people, making them vital and powerful. As Fabry's delightful time line tells us, ancient Roman epileptics drank the blood from gladiators in light of this freighted belief. It wasn't just to heal seizures but also to become physically stronger, more energetic. A thousand years later, and in a similar vein, Renaissance priest Marsilio Ficino suggested the elderly could experience rejuvenation by drinking younger males' blood (no gladiatorial experience required). A German physician three hundred years later recommended skipping fluids altogether. Seniors should simply lie next to young females, not for sex but for some mysterious passive transfer of youthful vitality.

None of this has worked. Nobody alive today was also alive hundreds of years ago. That hasn't stopped contemporary scientists from exploring the basic idea that youthful bodies have something that senior bodies do not. If that something could be isolated, then added back, perhaps youthfulness could be restored to the elderly.

This approach turns out to have some scientific merit, at least theoretically. Early hints come from an experimental technique called parabiosis. It's the process of surgically hooking up the vasculature of two creatures to each other. Slice away a bit of skin on each, stitch the exposed sections to each other, and their capillaries will connect as the wound heals. They then share their blood in real time. The geroscience version of parabiosis connects an old and young animal together, then studies what happens to the old guy. It's not conceptually much different from Ficino's ideas.

These experiments have been done, and it appears the old priest was on to something. The muscles of older mice get stronger and their hearts get healthier. Nearly every organ measured, including the brain, shows positive changes.

One of the best-known brain parabiosis experiments—famous because it worked—comes from the laboratory of Tony Wyss-Coray at Stanford. After joining mice in pairs and letting their circulatory systems mingle for a while, he observed dramatic changes in both structure and function in the old animals' noggins. Throughout the hippocampus, he saw both increased dendritic density and synaptic plasticity. Wyss-Coray's lab then went after the secret sauce, found that it was the plasma of the donor, and injected old mice with the plasma of young mice. They saw youthful alterations in the old mice's ability to learn things, with transformations in memory skills, spatial abilities, and fear-conditioning responses. These mice, it seemed to Wyss-Coray, were getting younger. He wrote in a paper published in *Nature Medicine*: "Here we report that exposure of an aged animal to young blood can counteract and reverse preexisting effects of brain aging at the molecular, structural, functional, and cognitive level."

That's quite a thing to say. Wyss-Coray interpreted these experiments as "restarting the aging clock," and he's not shy about using the word "rejuvenation" to describe its success. His enthusiasm led to the formation of a clinical trial in humans, injecting Alzheimer's patients with young plasma. Having finished these first experiments, the lab is now evaluating the results.

Skepticism being the reserve currency of the scientific world, not everyone is enthusiastic about his interpretations. Amy Wagers, a Harvard scientist who's done similar age-related parabiotic work, thinks rejuvenation is a bridge too far. "We're not de-aging animals," she explained in an interview in *Nature*, "we're restoring function." She believes young blood is simply assisting elderly repair systems to up their game. As we discussed, these systems invariably fall down on the job as we get older, shouldering the responsibility for the most difficult parts of aging.

No exit, but a smoother ride

From genes to drugs to blood swapping, what do we make of all these efforts? No question these scientific advances are nothing short of amazing. But amazing in the lab world and practical in the real world are two different things. Currently, we don't know enough to be optimistic about whether we can find a Fountain of Youth. The data aren't trending well, and given the complexity of the issues involved, it may be a long time until they do. From a research perspective, two different issues are being addressed—longevity versus aging—and neither one is getting us to immortality.

Take the gene work on longevity. Researchers have found great success with lengthening life in lab animals. In humans, though, what we're mostly left with isn't long life. It's cancer.

Much of the pharmaceutical work—and probably all of the parabiosis studies—concern aging, a different process. Ameliorating damage from randomly dysfunctional repair systems will certainly make our latter years more comfortable. It might even cure Alzheimer's. But it won't make us live forever. Death still has no clearly marked exits. Ultimately, sadly, compellingly, this grim highway is where the Immortality Bus is headed.

That doesn't mean we're entirely without hope, of course, or optimism about the process of aging. I can state unequivocally that there has never been a better time in human history to grow old. And, as you've been reading, we can do much to make the transit through aging as smooth as possible. It's to that hope and that optimism that we turn next, and finally.

We're going to explore what an ideal day might look like during retirement—and how beautifully it matches with the day-to-day lives of the people who enjoy the most days on the planet.

SUMMARY

You can't live forever, at least not yet

- Aging is not a disease, rather a natural process. People don't die of old age; they die of biological processes that break down.
- Genetics is responsible for between 25 percent and 33 percent of the variance in life expectancy.
- The Hayflick limit is the threshold beyond which a cell can no longer divide, leading the cell to deterioration and, eventually, death.

your retirement

brain rule

Never retire, and be sure to reminisce

The idea is to die young as late as possible.
—Ashley Montagu

Things ain't what they used to be and never were.
—Will Rogers

THE MOVIE *Cocoon* has an interesting take on aging, kind of an alien-meets-nursing-home mash-up of a plot (I can just imagine the studio pitch meeting). Directed by former child star Ron Howard, the movie achieved both commercial and critical success. It garnered two Academy Awards, including one for best supporting actor.

The film starts with three older gentlemen in swim trunks walking through their nursing home. We see the usual stereotypes: residents in wheelchairs, people shuffling with walkers, an exercise class for the ambulatory, men and women with vacant stares. The trio walks past a bedridden man experiencing a medical crisis—emergency health team frantically responding, barking instructions amid a flurry of fluids and tubes—and out the door.

The men are sneaking into a swimming pool next door, which, we will discover, has the mysterious ability to make them feel young and vital again. After a swim or two, they begin acting as if they'd just IV'd a can of Red Bull. But this is more than a psychological lift. One swimmer eventually finds his vision improving enough that he resumes driving. Another swimmer is miraculously cured of cancer. The sentimental heartbeat of this movie is the transformation of these older folks—and their gratitude for this new lease on life. Though aliens are involved (What mid-1980s movie didn't involve aliens?), the movie dealt with a rare theme in Hollywood: what it's like to grow old.

The transformations in this movie remind me of the story that began this book. Do you remember the remarkable counterclockwise study of Ellen Langer? It involved a monastery rather than a pool, and it seemed to have *Cocoon*-like effects on the vitality of the men who participated. You might recall I mentioned that the book you're currently reading is all about what happened to those men, and it is time to explain what I mean.

How should senior citizens design their days? We now have most of the ingredients for outlining the way brain science answers that question. In this chapter, we're going to talk about that design, focusing specifically on what you should do when you retire. If you retire. We may not get to the extraordinary achievements of aliens visiting a nursing home. Yet we can do much better than simply sitting in an isolated building with vacant stares.

The worst thing you can do to yourself?

What is the ideal age for retirement? Don't look to Charles Eugster for an example. This athlete, born in 1919, was still steaming through life like a runaway locomotive until the age of ninety-seven. "Retirement is one of the worst things that you can do to yourself!" he once exclaimed.

Charles Eugster looked like a quintessential British general: royal bearing, great vocabulary, bad teeth. That last bit is puzzling, because Charles Eugster was a retired dental surgeon.

He was also a legend in the world of geriatric fitness. Eugster held senior division track records in the sixty-, hundred-, and two-hundred-meter races. He won forty gold medals in World Rowing Masters

Regattas. He took the senior World Fitness Championship four times. If you look online for pictures of him, you'll see him running, boxing, and lifting weights, his toothy grin shining like a lighthouse on the way to tomorrow.

Eugster is no friend of retirement, an enmity he considers critical to his success. "If you refer to the Queen of England," he once explained, "she has a terrific schedule. She is not somebody who jogs in the park of Buckingham Palace, but she does an enormous amount of standing. She is not someone who sits, and sitting is not healthy. The most important thing is that she has a job."

The brain scientists in the room would applaud. People envision retirement as filled with carefree living, lengthy travel, and finally getting to do what you've always wanted to do. In reality, retirees' carefree attitude is term limited. You feel a sense of "getting out of jail free" for a while, but the negatives soon creep in. The truth about retirement's sunny reputation?

It's a myth.

We now know that retirement is extremely stressful for most people. Of the forty-three top life stressors in the canonical Holmes-Rahe Life Stress Inventory, retirement comes in at No. 10, just above "major change in the health or behavior of a family member." The proof? Get ready for a fusillade of statistics, for the concept of terminal retirement takes fire from both physical health and mental health perspectives. These data are necessarily associative, but their sum total shoots down the myth and creates a stark choice. Retirement, it turns out, increases your probability of dying.

If you choose not to retire, you lower your mortality risk by 11 percent—and thus increase your probability of living.

Retirement by the numbers

Researchers have known for years that retirees tend to be in worse physical health than people of the same age who stay on the job. They're 40 percent more likely to have had a cardiovascular incident, like heart attack or stroke. Blood pressure, cholesterol, and body mass index all rise to unhealthy levels.

And it's not just cardiovascular threats. Retirees are more likely to get cancer, too. They have a greater probability of acquiring diabetes. Retirees are more likely to have problems getting around, because they're in greater danger of getting arthritis. The overall risk for *any* chronic health condition is 21 percent for seniors who stay in retirement. It's about half that for seniors who stay employed.

Mental abilities also slide downhill. People who retire experience a rapid decline in fluid intelligence scores compared with working colleagues. You recall that fluid intelligence is the ability to "flexibly generate, transform, and manipulate new information." This decline isn't small. Those who retire perform half as well on tests as those who don't. Overall memory scores are about 25 percent lower. Retirement is like writing the obituary of someone who hasn't died yet.

The risks for mental dysfunction—the psychopathologies—also march in this depressing statistical parade. Retirement increases the probability of acquiring a major depressive disorder by a whopping 40 percent. There's an uptick in the risk for dementia of any kind. If you retire at sixty-five rather than at sixty, you drop your risk for acquiring dementia by 15 percent. We even know the rate. For every year you choose to work after age sixty, your risk for dementia goes down 3.2 percent.

Bottom line? Research science has a simple, one-word response for the ideal age at which you should retire.

That word is "never."

That's a potent thing to say, but in real life, one size doesn't fit all. Personal circumstances vary in issues ranging from finances to family proximity. Not everybody will be physically strong enough to have a vigorous non-retirement retirement. And not everybody will want it. The data are compelling enough to provide broad suggestions, but these suggestions are not guarantees. Rather, following them puts the statistical winds at your back instead of in your face.

The good old days

Before we tackle an hour-by-hour plan for aging well, I want to talk about Kentucky Fried Chicken. I get a touch of nostalgia whenever I see pictures of those old rotating buckets with a few chickens still stuck on gibbets outside KFC restaurants. My mother and I used to frequent KFCs, back in the days when the Colonel Harland Sanders was still alive and, having sold the company, fuming about how awful the product had become. He called the extra-crispy recipe "a damn fried doughball stuck on some chicken." To put it mildly, Sanders had a colorful past. He sold tires, bought a hotel, established a ferryboat company, painted barns, ran through several marriages, and was involved in a gunfight in which someone actually got killed.

Most of his success occurred after he was old enough to draw Social Security, however—an example of the great power of not retiring. He sold his first franchise in 1952, at age sixty-two. He continued to market his product over the next decade, watching his ideas blossom into a corporation with hundreds of restaurants. He sold it in 1964 for millions of dollars to a future governor of Kentucky, then spent the rest of his days as a fussy spokesperson for his food. He died at age ninety.

That's quite a non-retirement. It all comes back to me in memory burps every time I see high-altitude rotating plastic buckets of chicken.

The colonel's story holds at least two not-so-secret ingredients for anyone interested in living a long life (perhaps you want to stay away from gunfights, though). The first is work, which provides a purpose in life, a routine, and a social web 25 percent larger than that of retired colleagues. The second is the life-giving power of nostalgia.

Most advertising professionals, pop culture gurus, and historians understand the unstoppable power of "the good old days." Yet they might be surprised to know that solid brain science says they're doing us a favor: nostalgic experiences have many cognitive benefits. Working mostly in the United Kingdom, social psychologists such as

Constantine Sedikides and Tim Wildschut have broadened the field's understanding of how rose-colored memories of the past affect the less rosy experiences of the present.

Sedikides and Wildschut define nostalgia the way the 1998 version of *The New Oxford Dictionary of English* did: "a sentimental longing or wistful affection for the past." But they don't measure it as English majors would. They developed a psychometric test called the Southampton Nostalgia Scale to assess how much nostalgia a person is experiencing at any given time. And a research instrument—the event reflection task—powerful enough to induce nostalgia experimentally.

Nostalgia is often characterized as a form of cognitive quicksand. Wallow too much in it, and a person can get stuck in the past. (The word "nostalgia" literally means "homecoming pain," because it was thought that the physical and mental problems experienced by medieval soldiers came from a "toxic" longing to return home.)

So what the researchers found was unexpected: nostalgia is actually good for you. We now know that people who regularly experience nostalgic stimuli are psychologically healthier than those who don't. And we even know why, both at the behavioral level and, astonishingly, at the cellular and molecular levels, too.

It is to these ideas we turn next.

The power of "our song"

Like many couples, my wife and I share an "our song" song, a tune that reminds us of our dating years. It's appropriately called "Reminiscing," by the Little River Band. The song is all about a couple nostalgic for some old tune reminding them of *their* relationship:

> *Now as the years roll on*
> *Each time we hear our favorite song*
> *The memories come along*
> *Older times we're missing*
> *Spending the hours reminiscing*

Every time we hear this song—it's elevator music these days—we stop and smile and give each other a kiss, occasionally teary-eyed. Call it the "Our Song Syndrome." We've been married more than thirty-five years as of this writing. They've been the happiest years of my life.

Where does nostalgia gain such power, how does it work in the brain, and what does that have to do with retirement planning? Nostalgia has been a topic of growing interest to the scientific community, perhaps because we're all getting older. Nostalgia promotes something called self-continuity, linking who you were in the past with who you are now (in technical terms, a form of temporal self-stability in which autobiographical memory traces are integrated with present-day experiences). Here's the sequence of events that researchers discovered: (1) you wax nostalgic, (2) your self-continuity scores go up, and (3) good things happen to your brain. What kind of good things?

1. Nostalgia boosts "social connectedness" scores.

Social connectedness is defined as the subjective feeling of belonging to something or some group (like a tribe, or the Elks club, or the Greatest Generation) and being accepted by the other members.

2. Eudaimonic well-being increases.

This difficult word means "the sense of fulfillment that arises from achieving one's full potential as a human being." That might sound a bit squishy (What *exactly* is one's "full potential"?), but the experience has psychiatric consequences. The more "eudaimonia" you feel, the less likely you are to suffer from mood disorders. Eudaimonic well-being functions like garlic against the vampires of major depression.

3. Positive memories take priority.

Though waxing nostalgic is often described as "bittersweet," research shows you'll experience far more sweet than bitter. The positive priority is so robust it even shows up in brain scans.

These three attitudinal boosts are played out in the most practical corners of daily living. People who regularly experience the benefits of nostalgia are less afraid of dying. Long-term partners become emotionally closer when reminiscing about shared memories (the "Our Song Syndrome"). People become more generous to strangers after spending quality time in their "nostalgia zones." They also become more tolerant of outsiders, especially ones with perceived social differences. Even sensory information gets into the act. People placed in a cold room who begin experiencing nostalgia start feeling warmer, even though no one turned up the temperature.

The brains behind nostalgia

When researchers look in the brain via noninvasive imaging, they discover how—and why—nostalgia works its behavioral alchemy.

As people reminisce, certain memory systems kick into overdrive, mostly involving the hippocampus. That result is not surprising, on the level of discovering that cows make milk, for the hippocampus is involved in most of the brain's memory systems.

But more than memory is activated with nostalgia. Researchers discovered that regions like the substantia nigra light up like the Fourth of July during nostalgic experiences. So does the ventral tegmental area. Both regions are involved in generating the feelings of reward. Both use the neurotransmitter dopamine to make it happen.

This stimulation pattern has two interesting implications. First, your brain gives you a reward when you reminisce, so you want to repeat it. Second, reminiscing activates a neurotransmitter that is involved in learning and motor function, not just reward—and is, unhelpfully, fading with age.

All of a sudden we have a clue to the inner mainsprings of Langer's counterclockwise experiment. In it, waxing nostalgic didn't affect just the participants' attitudes. It also affected their bodies. Recall that the test subjects' eyesight improved. They played touch football. Since dopamine affects not only brains but also motor function (destruction of the substantia nigra results in Parkinson's, after all), it appears that the stimulation of dopamine in specific areas of the brain is the mechanism behind all this positivity. That's what nostalgia is good at, and since most senior brains exist in a critical dopamine drought, we now have very good news. Dopamine, as you know, is an extremely useful neurotransmitter to keep around, for both body and brain.

The bottom line: wax away. How far back should you set the way-back machine? And what kinds of memories are best to re-experience once you've arrived? Obviously, the more details you remember of your past, the more data points you have to feed the nostalgia beast. So what do seniors remember with the most clarity? That's up next.

The golden age: our twenties

Toy Story 3 involves a scene nearly unwatchable for my wife and me. It concerns Andy, the boy whose toys have been the subject of the previous movies. He's now grown up and going off to college. Too old for his childhood, too big for his toys, he's sorted his playthings into boxes, clearing out his room. Near the movie's end, just before he leaves home, Andy and his mother enter his mostly barren room. Mom suddenly stops. She looks around, eyes moistening, brain starting to defocus, mind suddenly abandoned to the thick fog of memory. Beholding a room no longer her son's, she clutches her throat, valiantly forcing back tears. Andy tries comforting her: "Mom, it's okay." She whispers, "I know. It's just . . . I wish I could *always* be with you." She turns suddenly and gives her son a heartbreaking hug.

The reason this is hard on us is that fictional son Andy is roughly the same age as our nonfictional son Joshua, who left for college in a similar fashion. I'm here to tell you the movie is spot-on. There are times when you wish your eyes came with windshield wipers.

Like most people, Josh is experiencing college in his late teens/early twenties. Those are important ages for geroscientists (yes, they study twenty-year-olds). Their research efforts uncovered a powerful ingredient for you to consider adding to your retirement planning.

The phenomenon requires looking at the gross domestic product of memory output over a lifetime of experience. If you ask a group of eightysomethings what items or events or experiences they remember the most, you quickly find two things: (1) retrieval is *not* an even experience and (2) you get the exact same retrieval response curve. The graph looks like an unfinished drawing of a double-humped camel. It's measuring a retrieval system that involves autobiographical memory.

This camel graph starts at zero and stays there for a bit, since nobody remembers much before the age of two or three. Retrieval climbs very quickly however, reaching its peak by age twenty. That peak composes the top of the first hump. Retrieval starts declining after age twenty-five, rapidly descending by thirty, and flatlining by fifty-five or so. That flat line is the saddle between the humps. Then recall begins climbing again, weakly, reaching a much smaller second peak (about half the size of the first) by age seventy-five. That's the second hump. What you get is an unfinished rendering of the profile of a Bactrian camel.

These humps are so persistent that scientists have named them. The weak one is termed "the recency effect," showing we remember newer events better than older ones. The larger first hump shows a clear retrieval bias around age twenty, skewed for events in our late adolescence/midtwenties. That isn't at all easy for scientists to explain (remember, scientists were quizzing *eighty*-year-olds). It too has a name—"the reminiscence bump." The phenomenon behind this bump is called retrieval bias.

A more pleasant way to get at retrieval bias is with a simple question: When did you have the most meaningful experiences of your long life? Though the question is fairly subjective (What does "meaningful" actually *mean*?), clear findings emerged. If you ask professional writers in their golden years at what age they read the books that changed their lives, you'll get a consistent answer: 75 percent will have read the most significant ones by age twenty-three. If you ask other seniors what's the best popular music they ever listened to, the ones defining "their generation," the answer is similar: the music they heard between the ages of fifteen and twenty-five. If you ask seniors what movies defined "their era," they routinely mention movies they saw—you can guess this by now—in their twenties. The most important political events? The ones that happened in their midtwenties. Ditto for social events. This is true not just with American seniors. Such biased retrieval is observable all over the world.

My peak reminiscence bump occurs in 1976, the year that saw the death of Mao Zedong and the birth of Reese Witherspoon. I remember the year as if it were yesterday, which, apparently, my brain still thinks it is. Having just gotten my driver's license, I recall gas costing less than a dollar a gallon (it was fifty-nine cents!). Average cost of a movie was about two dollars. A four-bedroom house in the Midwest was $36,500, wondrously, and the average yearly income in the United States was about $9,000.

The year was made more indelible because America was honoring its two hundredth anniversary of bicameral schmoozing. This bicentennial was celebrated with the writing of many history books, including *1876*, a best seller by Gore Vidal. Other authors had best sellers that year, too, including Agatha Christie (*Curtain*) and Leon Uris (*Trinity*).

Popular music was alive, stoned, and rocking in 1976. And universally quantifiable by the dulcet tones of Casey Kasem and his keep-reaching-for-the-stars radio show, *American Top 40*. Disco started to raise its cheesy-as-nachos hyper-head on the charts, though not without a fight: the best-selling single of 1976 was "Silly Love Songs," the decidedly un-disco-like hit by Paul McCartney and Wings.

The founding member of the *Rocky* film family was released in 1976. *One Flew Over the Cuckoo's Nest* was newly out. Both were appropriate, as this was an election year, where Americans chose their thirty-ninth president, Jimmy Carter. That was hardly the year's only history-changing event. Several months prior to the election, in April, a little company called Apple became a corporation.

Quite a year, 1976. Reminiscing about it is welcome relief, for it gives me something to think about other than our kids leaving for college.

Thawing out in our sixties

Besides reminiscence bumps and thinking the-best-darn-music-occurred-when-I-was-in-high-school, senior brains experience something of a mystery.

Starting in our early sixties, for reasons nobody understands, certain memories from our past begin bubbling to the surface. It might be an old teacher's face. It might be a junior high dance, a commercial jingle, the smells of a Woolworth's department store.

These memories aren't just fragmented shards from our glittering past. They're full-blown memory traces with specific identifiers. They're remote, containing contents you've not consciously considered in decades. They're also startlingly clear, as if they really *had* occurred only yesterday. And they're almost always memories laid down in the reminiscence bump. Scientists call these volatile items "permastore memories," a play on the concept of the permafrost. A better term might be "permathaw," for the brain appears to be defrosting the strata of memories laid down the same year you wondered how you were going to pay for college.

Several independent lines of research point like giant flashing neon fingers to one short time in your life. From bumps to books to permastores, your brain really favors experiences from your late teens/early twenties.

Caveats exist. Some research teams now place the reminiscence bump closer to age eighteen. People who've had extremely disruptive events in their lives (like migrating to a new country) produce retrieval biases in favor of the transitions, not a certain age. There may be sex-related differences, too, with the retrieval peak spiking at earlier ages for females, and at tighter, more focused time frames. These don't change the initial findings, of course, nor do they affect the general aim of the giant flashing neon fingers. In a hopelessly elliptical comment, we tend to remember memorable events, and what an untraumatized brain thinks was most memorable were the things you did in late high school and early college.

The newly young ones

Langer's counterclockwise findings were turned into a hit British reality television show called *The Young Ones*, tailored to the fish-and-chips set. It won a 2011 BAFTA Award, the British equivalent of an Emmy.

The producers persuaded six iconic British celebrities of a certain age (average eighty-one) to spend a week in the Langer time machine—all while being filmed. They re-experienced the target year of 1975, in a country house furnished with an "Aladdin's cave of seventies-ness." The immersion included political and popular culture; Margaret Thatcher had just been elected leader of the opposition, the Bay City Rollers were climbing the charts, and Arthur Ashe had just become the first African American to reach the Wimbledon finals. No cell phone, no Internet, no Brexit; the participants were completely insulated from the noisy world of twenty-first-century England.

Did it work? One participant soon felt healthy enough to put on his socks without assistance, roommates cheering him on. "It's like being in the land of the living," he said. Award-winning actress Sylvia Syms declared: "When I came in, I was in considerable pain. My back was painful all the time. I could barely walk. Now for some obscure reason, I cannot tell you why, that has improved. I also know that my trousers are looser!" Addressing one of her housemates, eighty-eight-year-old fellow actor Liz Smith, she continued: "[To] see that you lost some of your fear of walking without your stick is tremendous. That is a joy for all of us, I think." Another celebrity said he felt like "a new man."

This was a TV show, of course, not video evidence supporting a publishable paper. Aside from the interviews, there was no real attempt to measure improvement. Langer's work was far more serious. She did pre- and post-tests on motor skills, sensory discrimination tasks, and cognition. She also did comparisons to matched controls who were not experiencing the time warp.

The key turned out to be multisensory immersion, as if researchers had put a hand on the backs of seniors and gently pushed them into the past. Langer's participants were asked in advance to start discussing subjects related to their target year, 1959. The van transporting them to the monastery broadcasted 1959-era popular music over its "radio," faithfully interrupting with period-related advertisements. Upon disembarking, participants carried their own suitcases up to their rooms, no assistance allowed. Magazines and other props from 1959 were waiting. Group interactions, offered daily, involved discussions of relevant events of the late 1950s. At night they saw popular 1959 movies (*Anatomy of a Murder*) or had recreation nights replicating game shows (*The Price Is Right*).

Langer's results, though more quantitative, are sung in the same key as the British commentaries. Hearing scores improved in the experimental group, quantified by threshold sensitivities at 1000 and 6000 hertz. Near-point vision got better in the 1959 visitors, too, especially in the right eye. Finger length, a measure of manual dexterity, increased for more than a third (37 percent) of her experimental participants. In the control groups, only one person saw an increase, while a third suffered a decrease. Global physical measures, from posture to weight, also improved, as did performance on a whole-body dexterity test. One guy actually threw his cane away.

The tests weren't solely about senses and strengths. Cognitive tests contributed to the before-and-after picture, too. Assays included the digit symbol substitution test, a rigorous timed test of processing speed and memory. The experimental groups' post-time-warp scores were 23 percent higher than the scores of controls. Of controls, 56 percent showed a decline on their digit symbol test, compared with 25 percent of the experimental group. Clearly in evidence: improvements over baseline scores or a slowing of decline compared with controls. As with any research project, pesky caveats apply. The sample size was small, the length of time was short, and not all tests showed clear-cut victories. Hardly enough to wilt the conclusions, but these demonstrated the results were more useful as flashlights—illuminating areas for further research. Which did occur, causing Langer to conclude years later: "When the results of this study are taken in conjunction with the many findings of our research that were cited earlier, we feel there is enough evidence to suggest that the 'inevitable' decay of the aging human body may, in fact, be reversed through psychological intervention."

Quite a thing to say for one of Harvard's longest-tenured professors. Put all of that together and you have a powerful, unique ingredient to include in your retirement portfolio. But how? We have to live in the present for the most part, even if we shouldn't always dwell there. What might that look like practically? A Beatles song lights the way.

A day in the life (past)

As a young baby boomer, I admit to a fond appetite for Beatles songs. They weren't my primary source of musical nutrition growing up (I'm into an earlier generation of shaggy-maned musicians). Yet the first time I heard the song "A Day in the Life," I realized the nineteenth century didn't have a lock on long-haired musical genius.

As you may know, "A Day in the Life" is really two songs sutured together, the haunting first and third sections written by John Lennon. Lennon says his lyrics were inspired from newspaper articles he glanced over at the time ("I read the news today, oh boy"), several from the January 17, 1967, edition of the *Daily Mail*. The car wreck concerned the death of Guinness heir Tara Browne. The four thousand holes referred to an article about the dismal road conditions in Blackburn, a British city in Lancashire county. It may not lead you to write a hit song, but it's worth going hunting for newspaper editions from your youth, too. Then start making a collection of memorabilia from those years until you have a room's worth of stuff.

Call it a "reminiscence room."

Contemplate filling an area of your current living environment with nostalgic items—the ones most likely to evince strong dopaminergic reactions. This might include pictures of families and friends. It might include objects and posters of meaningful events. The sound system in the room might have easy access to the recordings of the Beatles or Beethoven or whatever music coaxes your strongest feelings of yesteryear. There'd be a television—perhaps an old device hooked up to new technology—devoted to looking at vintage TV shows and a collection of old movies. Finally, you'd display books popular in your time, either previously read titles or all those titles you swore you'd get around to. Rather than shying away from the past, a regular part of your day would involve celebrating it. This room is like your personal Fountain of Youth.

What years should you emphasize? If we compare the reminiscence bump against Langer's data, we run into inspirations, contradictions, and the great unknown. Reflexively, you might argue that the nostalgia should come hurtling into the present from the calculated reminiscence bump of your past. But you'll notice that Langer pulled from events the subjects experienced when they were in their late forties or early fifties, not their twenties.

Why didn't Langer use the reminiscence bump data instead? She didn't have a real time machine: reminiscence data didn't reach the literature until the mid-1990s, and Langer did her work in the early '80s. Does nostalgia exert such a widespread arc that you can still get wet from the fountain even if you aren't splashing around the reminiscence bump? Would Langer have obtained more powerful results had she wound the clock back a few more decades? Given that points of stimulation are much more numerous and available in the reminiscence bump, it's a reasonable experiment to try. Until we do, my recommendation is an informed suggestion, not a peer reviewed prescription.

A day in the life (present)

The Beatles song is inspiring to me for another reason: designing a present-day day in the life. What might your typical day look like, hour by hour, if long life and maximum cognitive health were your goals? What would you eat? Whom would you see? What would you do?

I'm going to imagine a seventeen-hour period in the life of one senior. Her name is Helen. She's a seventy-year-old retired teacher whose husband died a year ago. She is able to get around, a bit fragile (arthritis), but in otherwise good health and able to drive. She's living alone in a two-bedroom apartment, with grown children nearby. Here's what a typical day might look like if Helen implemented many of the suggestions outlined in this book.

Again, please note these recommendations are aspirational, not prescriptive. Research shows that millions of people live relatively healthy lives far past their seventies, like Helen. But everyone's living situation is different. Also think of Helen's daily schedule like a buffet table. You can mix, match, and modify at your pleasure—whatever fits your style, energy levels, work and family situations—and still reap great benefits. Ultimately, this schedule is as individual as you and your journey through the aging process.

7:00 a.m.

Helen wakes up, reads a note she's kept by her nightstand, and breaks into a smile. Breakfast consists of berries, whole-grain cereal, and nuts, washed down with a refreshing fifteen-minute meditation. She focuses on a short body scan, the sine qua non of mindfulness, before planning her day.

Helen does this because she's concerned about the stress in her life, both now and in the future. The breakfast is high-octane MIND diet, the nutritional program shown to reduce the incidence of Alzheimer's. She's been eating it for a while, and every bite helps allay her concerns about her future mind. Mindfulness reduces this stress; improvements in her cardiovascular system are already apparent. She's also sleeping better, and oddly enough, her vision has improved. These improvements go a long way toward increasing the mileage Helen will have with her grandkids. Like a smooth-running, high-end transmission, she's now ready to switch her morning into high gear.

8:00 a.m.

It begins with a knock on her door. It's Helen's walking group, a wonderful assortment of dear friends. They call themselves the Galloping Grannies. They're going to take a brisk, thirty-minute walk around the block, something they do several times a week. One of the women has recently been widowed, and Helen has been a godsend to her, walking and talking her through her grieving each morning.

Helen puts this activity in the pole position for many reasons. Certainly, exercise improves her executive function, a fact she feels every time she balances her checkbook or thinks about her finances. She also connects with old friends, some just beginning to experience aging's more life-jarring disruptions. These interactions work like medicine. It's the first of many social events she'll experience that day, each one good for body and soul. Lovely thing, she muses, to have a brain vitamin wrapped in the safe folds of friendship.

9:00 a.m.

After saying goodbye, Helen begins what she calls her "education time." She's taking two classes (alternating days) at the local community college. Today it's music class, consisting of theory and piano lessons. Tomorrow will be French lessons. She's always wanted to learn French because she's always wanted to take a trip to Paris. She's going next summer. Knowing that aging doesn't grade on a curve, she is anxious to get this started while she's still healthy enough to travel.

The second part of her "education time" involves volunteering as a teacher in the community college's ESL (English as a Second Language) program. There are immigrants of all ages in the class, a few as old as she. Even so, Helen often feels parental, a lifeline to people who find English bewildering, American culture perplexing, and friends to whom they can speak rare.

Helen was as strategic as Napoleon about how she structured her "education time." Because she speaks no French, the class forces her brain to become immersed in topics completely foreign to her. Such challenges slow down general cognitive decline. They're also boosters of episodic memory (memory for events) and working memory (short-term memory). Her ESL class is also good for her brain, acting like a roll cage against the rough-and-tumble corners of aging, mostly because it forces her to take other people's viewpoints. Her students aren't people who share her culture. The class is also intergenerational, filled with young parents, teenagers, even a grandfather. To teach them effectively, she'll have to acclimate to their unique perspectives. Such exercises keep depression in last place in her life, reduce stress, and increase her chances of living longer.

Helen deliberately chose ESL as *voluntary* charity work. It allowed her to be part of something "larger than she is," an activity proven to create and sustain positive worldviews. It's not lost on her that these classes represent more socializing interactions, another handful of brain vitamins. The only thing the classes have in common is that the people in them all know her.

Noon

Helen comes home exhausted. And hungry. Lunch consists of a salad with olive oil, a lot of fruits and vegetables, and a little chicken. She takes a quick nap, thirty minutes tops, before beginning the afternoon's activities. Helen's part of a book club, and today it's her turn to host. She prepares light snacks and begins reading, the first of two books she'll peruse that day.

She started the club. It always produces lively, occasionally intense discussions. She's always sorry to see her club-mates leave, even the ones with whom she regularly disagrees. Helen's as confidently opinionated as a political primary, and so is everyone else in the club.

Such friendly bantering is a blessing in disguise, for measured disagreement increases fluid intelligence scores. It makes Helen's brain more efficient, filling her cognitive reserve. After the meeting, her brain feels as if it had been lifting weights. The activity is important all by itself, however, for reading is a friend with Fountain of Youth benefits. A consistent habit of it lengthens life.

She's not done socializing. After cleaning up, Helen turns on her computer to immerse herself in the brave new world (to her) of social media. This consists mostly of Facebook acquaintances, and she visits the usual sites of friends and family. Her kids bought her a cell phone several years ago, and now it's a constant companion. Her daughter regularly sends texts with the latest pictures of the grandkids. Helen gets lost in the chat, tapping away with the gleeful enthusiasm of a teenager.

Then she does something really strange. Texting finished, and daughter gone, Helen turns to a video game (another gift from her kids). It's a brain-training exercise. She resisted it for a while (she'd heard mixed things about video gaming), but her kids gave her only thoroughly researched titles. Since the computer was already on, it was easy for her to click to a game involving car racing. Though she still doesn't like it much, she's getting surprisingly good at it. If she continues playing, her attentional states will improve rapidly, especially her ability to resist distractions. Her short-term memory will get another series of reps in the cognitive gym.

3:00 p.m.

After Facebook, and many laps around the virtual racetrack, Helen's ready to get moving again. She's been taking a ballroom dance class every afternoon. At first she found the class as obnoxious as tear gas. Not only did the close interactions remind Helen of her husband, but the required physical coordination with a stranger also proved difficult. Her attitude changed, happily, as the class progressed. Now she finds the synchronized human contact refreshing and surprisingly easy. She doesn't know this, but her balance is improving, as well as her posture, and her risk of falling is, well, falling. She's not attracted to any of the single men in the room, yet dancing seems to take the edge off the grief of losing her husband. It's her final social interaction of the day.

Returning from dance class, she notices it's 4:30 p.m.—about half an hour later than she'd like—but she still immediately thinks of bedtime. Helen's not headed off to an early slumber; she's simply beginning to prepare for sleep later that night. After late afternoon, there's to be no more caffeine, alcohol, exercise, or computers. That's so by 11:00, she'll be drowsy enough to enter delta-wave land.

At 5:00 p.m., she prepares dinner. Tonight it will be fish and pasta and lots of produce. In direct violation of her mandate not to drink alcohol after 5:00, she has a glass of red wine. Maybe next time she'll have it with lunch.

7:00 p.m.

Helen now gets ready to experience her favorite part of the day, which she's christened the "H. G. Wells Evening." She's going to step into a time machine, a room she's specially outfitted to re-experience the world of the mid- to late 1960s. There are posters on the wall, an old turntable on the desk, plenty of vinyl, a TV, a DVD, and a perfume bottle. The fragrance is Joy by Patou, which she wore while dating her late husband. She puts a dab on her wrist and cranks up the music, which is anything from the Beatles to Aretha Franklin.

Dessert consists of an Eskimo Pie, and she lunges into it with the delicacy of a shark. Ignoring the brain freeze, she picks out an old book, selecting something that reminds her of college days. She's currently rereading *Christy*, a novel by Catherine Marshall.

An hour into her reading, she catches a sniff of the perfume. Memories trickle into her mind, and soon tears trickle down her cheeks. It's helpful to watch a DVD of an old television program called *Laugh-In*, a sketch-comedy show popular in the late 1960s. She laughs so hard her tears change their emotional complexion, and now she cries for different reasons.

These H. G. Wellian exposures are deliberate. The time-warp room is filled with events experienced during Helen's reminiscence bump, and she has given it a full-frontal sensory assault: sights and sounds, tastes and smells. They're all designed to increase dopamine levels in her brain. And she did part of it absorbed in a book, pushing her daily grand reading total to a life-stretching figure of three-plus hours.

11:00 p.m.

After such a full day, Helen's run out of gas. She has one more task before she falls asleep (usually around midnight), and it involves pencil and paper.

Helen divides the paper into two columns. She writes in the first column three things that happened that day—things that made her smile or made her feel grateful. In the second column, she describes why the events gave her such feelings. Popular on this list are interactions with the grandkids, which make her feel connected. Another is the ability to still drive, and she's grateful for the independence. Helen has discovered that even on her crummiest days, there's still something to appreciate.

She puts the list on her nightstand, climbs into bed, and is soon fast asleep. Next morning, the first thing she'll read is that list. It will make her smile, as it always does. Then she'll be ready for another day, knowing she is doing all she can to change both the number and quality of her days.

She's decided to design her life according to brain science, and it's the best thing she ever did.

Building a mighty river

What lies behind this tale is an important idea: a multipronged strategy is the best approach for maintaining cognitive function. Is there empirical evidence that this approach works? Can you really rearrange the cognitive furniture inside your mind and make your life easier to live in? The answer appears to be yes, and exhibit No. 1 is a big, fat randomized trial by a group of Scandinavian researchers.

They wanted to know what would happen if seniors (sixty to seventy-seven years old) ate up a combo plate of diet, exercise, and brain-training programs. They christened the experiment FINGER, short for Finnish Geriatric Intervention Study to Prevent Cognitive Impairment and Disability (I think you have to speak Finnish to make the letters work). The men and women of the study, more than twenty-five hundred strong, were selected on the basis of an elevated risk for dementia. Then the researchers took the five-star approach to any behavioral study, randomly assigning the seniors into experimental and control groups.

For two years, the experimental group ate foods from the Mediterranean diet. Simultaneously, they submitted to a vigorous exercise program consisting of aerobics, strength training, and balancing drills (eventually reaching two or three sixty minute sessions per week). They played a smorgasbord of games addressing executive function, processing speed, and memory (they played in fifteen-minute bursts, two to three times per week). And to closely monitor their health, the experimental group frequently saw doctors, nurses, and allied health staff, each visit chock-full of cardio and various metabolic tests. The control groups got none of this gold-plated treatment. Aside from normal health monitoring, they simply received standard recommendations for health.

The results were impressive. Memory test scores improved 40 percent in the treatment group compared with the controls. Executive function improved 83 percent. Processing speed improved a whopping 150 percent. The controls either languished or got worse. In fact, the overall cognitive performance of the untreated cohort declined 30 percent.

Does it work to make many healthy lifestyle changes all at once? Yes, it does, and in almost every way you can measure it. Aging takes you closer to horizons that no longer fade into the distance, but you can journey to the vanishing point with a healthy brain, full of life and enthusiasm.

That brings us nearly full circle. We started this book with a remarkably sprightly David Attenborough describing his sojourn along the Amazon River. He said this mighty river became mighty not because it started life as a giant waterfall, cascading down some Olympian mountain. It began small, becoming grand because of the contributions of many small streams and rivulets gathering together, gaining momentum, creating an *e pluribus unum* out of the world's mightiest river.

Your design for your life is just like this. In paying attention to the individual streams—from socializing with friends and reducing your stress to staying physically active and practicing mindfulness—you can flow more smoothly through aging.

Take your cue from a motley crew, all living the longest lives on the planet.

Hot spots of healthy old age

You might be hard-pressed to find much in common with a fisherman in Okinawa, a pastor in Southern California, a hotel owner in Greece, and a farmer in Italy. But you can. That's what Dan Buettner found. Buettner, an explorer, holder of several endurance-cycling records, and best-selling author, is as handsome as a 1950s movie star. With his financial research tank filled with funds from the National Geographic Society and National Institute on Aging, Buettner double-teamed with Italian demographers to scour the world, looking for "hot spots" of longevity. They found five, scattered from southern

Okinawa to Southern California. These spots share the feature of being populated by people who live not just ridiculously long lives but ridiculously *healthy* long lives.

The findings are impressive. Fully 80 percent of the eighty-year-olds on the Greek island of Ikaría are still working—and still growing their own food. They have only 20 percent of the dementia rate of Americans. They live longer than their US counterparts by seven years.

There's a peninsula in Costa Rica where the probability of living to ninety is more than double the US figure. A sixty-year-old man on the peninsula has a chance of celebrating his hundredth birthday that's seven times that of a Japanese man the same age.

The list goes on. Female Seventh-day Adventists in Loma Linda, California, have a life expectancy of eighty-nine, a decade longer than their non-Adventist next-door neighbors. There's a mountaintop in Sardinia that's home to the world's highest concentration of men ages one hundred and older. There are places in Okinawa where the prevalence of female centenarians per capita is *thirty times* that in the United States. These women live the healthiest lives on the planet, right up until they die. Buettner christened the regions that contain these aging champs "Blue Zones," for the color of pen he used to create concentric rings around the original maps.

What in the world are Blue Zoners doing to live such a long time? People around them would certainly like to know, especially in the United States. One-fifth of all Americans over the age of sixty-five already have mild cognitive impairment, the first knock on the door of life-wrenching dementia. One-third of all Americans have high blood pressure, the first knock on the door of life-ending cardiovascular issues. What's frustrating about these deficits is that a great deal of our aging lives is under our control. Only a paltry 20 percent of our tenure on Earth is supervised by how well we picked our parents. That means that 80 percent of how long we live is up to us, or at least up to our environment. And that's just according to one study, generous with genes. More miserly research says we can blame only 6 percent of the variance on our genetics; a hefty 94 percent is soldered to our lifestyle.

In a 2012 *National Geographic* article, Buettner wrote about the secrets of the Blue Zoners. Two things pop out to me: they all made similar lifestyle choices, and nearly all of their choices match up with the cognitive neuroscience we've covered in this book. These people live in far-flung regions, inhabit vastly different cultures, and don't communicate much with the outside world. No scientist told them what to do. Yet they arrived at the same peer-reviewed spot, and each one enjoyed an extraordinarily long and healthy life.

Buettner and neuroscience agree on how they did it—and how we can do it, too.

Friendship

All Blue Zoners have active social lives, so Buettner told *National Geographic* readers to "keep socially engaged." He observed, "They put their families first." That should sound familiar. As discussed in our friendship chapter, the rate of cognitive decline is 70 percent less in seniors with active and highly interactive social lives. Benefits come as long as those interactions are positive and fulfilling. Not surprisingly, friends and family are the richest sources of the benefits, with stable marriages being particularly powerful. So are regular interactions with various age groups. Marriages and grandkids. Nothing could be more lively.

Stress

Brain science clearly confirms the obvious health benefits of reducing stress. Mindfulness training is a powerful way to get there. Mindful seniors have fewer infectious diseases, boast an 86 percent improvement in markers for cardiovascular health, and show a 30 percent improvement in attentional states. The same ideas show up in two remarkable suggestions from Buettner's article. "Observe the Sabbath," he wrote, describing how Seventh-day Adventists regularly push the pause button on their busy Southern California lives. This includes church and prayer and, much like mindfulness, a mandated calming break in routine.

Friends also buffer against the harmful effects of stress. This is reflected in Buettner's second suggestion: "Keep lifelong friends." As the song says, one is really the loneliest number. Lifelong friends are the antidote.

Happiness

Optimistic people live almost eight years longer than the glass-half-empties do. And they're more likely to experience what Martin Seligman calls "authentic happiness." One unjammed route to this happiness is to identify and pursue something that gives your life meaning. A belief in something—or someone—larger than you, giving to charity, doing some good in the world: all qualify. "Have faith," Buettner writes, again referencing the Adventists. And "find purpose," describing the wise advice of the Okinawans.

Memory

Keeping your mind active—whether reading or learning a new language (or engaging in what Denise Park calls "productive learning") —affects cognition. Reading more than 3.5 hours a day even extends one's life by a whopping 23 percent. Playing brain-training games that increase your processing speed also improves your working memory. (But if you don't like gaming, don't worry: most of the Blue Zoners stayed sharp past one hundred without ever playing *NeuroRacer*.)

Sleep

Brain science makes the obvious observation that sleeping well means minimizing stress. You can do this by having lots of social interactions (keeps depression at bay), maintaining a consistent schedule, and engaging in regular exercise. Blue Zoners are champions at all three. Many are involved in food-related jobs that pay close attention to the rhythms of the day. Their sleep habits have not been published, but their lifestyles already predict what the data would show.

Exercise

The brain science research is clear and unequivocal: exercise is good for you. Its effects on the life of the body are as robust as they are canonical, your cardiovascular system reaping the lion's share of rewards. But exercise also helps the life of the mind. Aerobic exercise has benefits ranging from better memories to better emotional regulation, boosting executive function by 30 percent.

None of this is lost on the Blue Zoners. Every one of them has an active lifestyle, in ways that sometimes boggle the mind. Buettner describes, for example, one morning in the life of a seventy-five-year-old farmer named Tonino. This Italian split wood, milked cows, slaughtered a calf, and escorted his herd of sheep through four miles of grass—all before 11:00 a.m. Buettner wrote simply, "Be active every day." That sentence is marinated in brain science.

Diet

Every Blue Zone group had something to say about diet, much of it aligned with the Mediterranean and MIND diets. These diets have been shown to improve memory, lessen the chances for stroke, and be robustly associated with long life. Buettner's words might easily have come from these peer-reviewed findings. "Eat fruits, vegetables, and whole grains," Buettner wrote, describing the diets he saw. "Eat nuts and beans," added the Adventists camp. Sardinians had the best recommendation in the lot: "Drink red wine" and "Eat Pecorino cheese." The Okinawans gave the hardest counsel. "Eat small portions," they advised. Brain science backs up all of them.

Retirement

Buettner's descriptions of Blue Zoners' day-to-day activities make it clear. *Most of them weren't retired.* Many senior Okinawans were still fishing (skin diving with nets!), many senior Adventists were still active in their charities, many Sardinians were still farming. And, of course, Tonino was still splitting wood and walking four miles before lunch. "I do the work," Tonino said. "My *ragazza* [gal] does the worrying."

Taken together, the congruency of the Blue Zone lifestyles with scientific findings is both extraordinary and expected. The people who live the longest lives on Earth show us something very hopeful. Though the house of death always wins, we can, for a while, play a surprisingly strong hand.

SUMMARY

Never retire, and be sure to reminisce

- People who retire from a job are at greater risk for physical and mental disabilities, including cardiovascular diseases, depression, and dementia.
- Nostalgia is good for you. People who regularly experience nostalgic stimuli are psychologically healthier than those who don't.
- Most seniors retrieve the clearest memories from their late teens/early twenties as well as from the most recent decade of their life.
- People who live in "Blue Zones," areas of the world where life expectancy is the longest, tend to be active, eat well, reduce stress, stay optimistic, and maintain a social life.

Now, Voyager

No matter how long we each have, it's inspiring to think about how our human story will continue to unfold. We've all seen so many remarkable things in our lives already. For me, science nerd that I am, one of the most amazing is the ongoing *Voyager* space program.

I first heard about the *Voyager* space program during an interview with legendary astrophysicist Carl Sagan. Launched in 1977, *Voyagers 1* and *2* were tasked with visiting the gas giants Saturn and Jupiter. The late Dr. Sagan described how gold-plated records had been placed on board the ships. The records were filled with terrestrial location information, pictures and sounds of Earth, and various artistic achievements, including a song by Chuck Berry ("Johnny B. Goode"). This literal record of human activity functioned as an interplanetary greeting card, just in case the spacecraft encountered intelligent life curious enough to want to know who mailed it.

I remember being gobsmacked, hearing Sagan's words as a kid. Planets! Scientists! *Aliens!* But this wasn't Hollywood; it was real stuff. My mind was positively electric. I was a wet-behind-the-years undergraduate in those days, contemplating with some trepidation whether I should commit to a career in science. It was different world then. A gallon of milk cost $1.68, a Honda Accord $4,000. Life expectancy (measured from birth) was around seventy-three years of age.

Three years later, *Voyager 1* arrived at Saturn. The ringed planet, ready for its close-up, did not disappoint. What pictures these tiny, hardy crafts could take! Like a celestial celebrity, Saturn made the cover of *Time* magazine, *National Geographic*, and countless scientific journals. The mission of *Voyager 2* was extended to see Neptune; it made its closest approach in 1989. More beautiful pictures, more magazine covers, a giant orb luminous as a Christmas light, blue as a sapphire.

I was rendered just as slack-jawed by the images of Neptune as I had been by Saturn years before, even though my life and world had changed considerably. I was now a postdoctoral fellow, the science career having won out, my minty-fresh PhD only a year old. A gallon of milk cost $2.34, a basic Honda Accord $12,000. Life expectancy was around seventy-five years of age. The future, it seemed, was as limitless as a universe.

The spaceships were still zipping along in 2012, their planetary flybys long past, but their merits hardly diminished at all. In August of that year, *Voyager 1* became the first human-designed craft to enter interstellar space. This tiny message-in-a-spacecraft, still tethered to its home world through a mere slender thread of electromagnetic radiation, hurtled outward into the heliospheric void. Most of its instruments had been shut down, but those that remained still bravely transmitted data.

And I *still* felt like a kid. This thrill remained, though nearly everything else since 1977 had changed. Now with a gray beard, a house full of teenagers, and publications and books and a lifetime of science and teaching under my belt, I felt as though my undergraduate years were as distant as *Voyager* to Earth. Milk was now four dollars, the Honda $24,000. Life expectancy was just south of eighty years. Yet as I read about the press release of this intrepid, interstellar spacefaring friend, my brain felt no loss of enthusiasm. Or function. It was still capable of loving life, of digesting information, of taking big perceptual bites out of this miraculous universe.

It still is.

So is yours. A sense of preserving wonder and curiosity—and the good news that you can still have both—are what I'd like to leave you with as I close this book. With love and care (and admittedly luck in the genetic roll of the dice), our brains will remain facile and flexible enough to keep our imaginations fertile regardless of age. It's never too late to embrace your friends, write down what you're grateful for, learn a language, learn to dance a jig, learn *anything*. You may have more years than you know. Aging always makes claims on a body, but not always on a mind.

Centuries after we're dead, the *Voyager* twins will keep chugging along, the little spaceships that could. And they will be ready to play Chuck Berry to whatever—or whomever—will listen.

After all these years, I still get the shivers.

references

Extensive, notated citations
at www.brainrules.net/references

10 brain rules for aging well

1.

Be a friend to others, and let others be a friend to you

2.

Cultivate an attitude of gratitude

3.

Mindfulness not only soothes but improves

4.

Remember, it's never too late to learn—or to teach

5.

Train your brain with video games

6.

Look for 10 signs before asking, "Do I have Alzheimer's?"

7.

MIND your meals and get moving

8.

For clear thinking, get enough (not too much) sleep

9.

You can't live forever, at least not yet

10.

Never retire, and be sure to reminisce

acknowledgments

THERE ARE FAR TOO many people to thank for making this project a reality, but here's a partial list. Thanks to cover designer Nick Johnson, copy editor Judy Burke, fact checker Erik Evenson, proofreaders Carrie Wicks and Nick Allison, and those who read early drafts, including Susan and Bob Simison, Carla Wall, and Vicky Warnock. Thanks also to the tireless efforts of the developmental editor of this book, Tracy Cutchlow, and publisher Mark Pearson.

Most of all to my family: My wife, Kari, the oxygen I breathe! And my two sons, Josh and Noah, for constantly showing me that curiosity is as old as the Pleistocene and as recent as last week.

about the author

DR. JOHN J. MEDINA is a developmental molecular biologist focused on the genes involved in human brain development and the genetics of psychiatric disorders. He has spent most of his professional life as a private research consultant, working primarily in the biotechnology and pharmaceutical industries on research related to mental health. Medina holds an affiliate faculty appointment at the University of Washington School of Medicine, in its Department of Bioengineering.

Medina was the founding director of two brain research institutes: the Brain Center for Applied Learning Research, at Seattle Pacific University, and the Talaris Research Institute, a nonprofit organization originally focused on how infants encode and process information.

In 2004, Medina was appointed to the rank of affiliate scholar at the National Academy of Engineering. He has been named Outstanding Faculty of the Year at the College of Engineering at the University of Washington; the Merrill Dow/Continuing

Medical Education National Teacher of the Year; and, twice, the Bioengineering Student Association Teacher of the Year. Medina has been a consultant to the Education Commission of the States and a regular speaker on the relationship between neurology and education.

Medina's books include *Brain Rules: 12 Principles for Surviving and Thriving at Work, Home, and School; Brain Rules for Baby: How to Raise a Smart and Happy Child from Zero to Five*; *The Genetic Inferno*; *The Clock of Ages*; *Depression: How It Happens, How It's Healed; What You Need to Know About Alzheimer's*; *The Outer Limits of Life; Uncovering the Mystery of AIDS*; and *Of Serotonin, Dopamine and Antipsychotic Medications*.

Medina has a lifelong fascination with how the mind reacts to and organizes information. As a husband and as a father of two boys, he has an interest in how the brain sciences might influence the way we teach our children. In addition to his research, consulting, and teaching, Medina speaks often to public officials, business and medical professionals, school boards, and nonprofit leaders.

index

F

N

O

P

T

U

V

W

Y